Hospitality and Tourism Education in China

This book is the first to systematically introduce China's tourism education system and the various tourism education practices in China to the international audience and stakeholders. China has the world's largest tourism education system, which consists of over 1,000 higher learning institutions with tourism-related programs and over half a million of tertiary-level students studying in these programs. Despite the industry scale, internationally, little is known about this tourism education system and how it operates. Knowledge and better understanding of China's tourism education system are important as tourism becomes one of the critical forces transforming economy, society and environment.

The book offers an historical evaluation of China tourism education development and elaborates on the current industry status and practices in different subject fields of China's tourism education, including tourism management, hospitality management, events and festival management in higher education, tourism vocational education, tour guides training and certification, master of tourism administration (MTA) education as a unique education model in China, PhD education in tourism, tourism curriculum, research and international collaboration in tourism education in China.

The book provides relevant knowledge to international tourism education providers, industry practitioners, human resource managers, government officials, and tourism academics, researchers, and students.

Jigang Bao is Professor in the School of Tourism Management and the Director of the Centre for Tourism Planning and Research at Sun Yat-sen University, China.

Songshan (Sam) Huang is Professor in the School of Business and Law at Edith Cowan University, Australia.

Routledge Advances in Management and Business Studies

Competition, Strategy, and Innovation
The Impact of Trends in Business and the Consumer World
Edited by Rafał Śliwiński and Łukasz Puślecki

Critical Perspectives on Innovation Management
The Bright and Dark Sides of Innovative Firms
Edited by Patryk Dziurski

Operations Management in Japan
The Efficiency of Japanese Manufacturing
Hiromichi Shibata

Stakeholder Management and Social Responsibility
Concepts, Approaches and Tools in the Covid Context
Ovidiu Nicolescu and Ciprian Nicolescu

Japanese Business Operations in an Uncertain World
Edited by Anshuman Khare, Nobutaka Odake and Hiroki Ishiruka

Entrepreneurship and Culture
The New Social Paradigm
Alf H. Walle

Hospitality and Tourism Education in China
Development, Issues, and Challenges
Edited by Jigang Bao and Songshan (Sam) Huang

For more information about this series, please visit: www.routledge.com/ Routledge-Advances-in-Management-and-Business-Studies/book-series/ SE0305

Hospitality and Tourism Education in China

Development, Issues, and Challenges

Edited by
Jigang Bao and
Songshan (Sam) Huang

Routledge
Taylor & Francis Group

LONDON AND NEW YORK

First published 2022
by Routledge
2 Park Square, Milton Park, Abingdon, Oxon OX14 4RN

and by Routledge
605 Third Avenue, New York, NY 10158

Routledge is an imprint of the Taylor & Francis Group, an informa business

© 2022 selection and editorial matter, Jigang Bao and Songshan (Sam) Huang; individual chapters, the contributors

The right of Jigang Bao and Songshan (Sam) Huang to be identified as the author of the editorial material, and of the authors for their individual chapters, has been asserted in accordance with sections 77 and 78 of the Copyright, Designs and Patents Act 1988.

British Library Cataloguing-in-Publication Data
A catalogue record for this book is available from the British Library

Library of Congress Cataloging-in-Publication Data
Names: Bao, Jigang, editor. | Huang, Songshan, editor.
Title: Hospitality and tourism education in China : development, issues, and challenges / Jigang Bao, Songshan Huang.
Description: First Edition. | New York : Routledge, 2022. |
Series: Routledge Advances in Management and Business Studies |
Includes bibliographical references and index. | Contents: Hospitality and Tourism Education in China: An Overview — Tourism Management Undergraduate Programs — China's Higher Education in Hospitality Management — Event Management Education in China — Tourism Vocational Education — Tour Guiding Education, Training and Administration — Master of Tourism Administration Education — Tourism PhD Programs in China — Curriculum Settings and Comparisons — International Collaboration in Tourism Higher Education — Tourism Research in China — Critical Issues, Challenges and Future Prospects.
Identifiers: LCCN 2021040367 (print) | LCCN 2021040368 (ebook)
Subjects: LCSH: Tourism—Study and teaching (Higher)—China. | Hospitality—China.
Classification: LCC G155.8.C6 H67 2022 (print) | LCC G155.8.C6 (ebook) | DDC 910.71/151—dc23
LC record available at https://lccn.loc.gov/2021040367
LC ebook record available at https://lccn.loc.gov/2021040368

ISBN: 9780367435707 (hbk)
ISBN: 9781032198972 (pbk)
ISBN: 9781003004363 (ebk)

DOI: 10.4324/9781003004363

Typeset in Galliard
by codeMantra

Contents

Figures

Tables

Contributors

Jigang Bao is a professor in the School of Tourism Management, Sun Yat-sen University, China.

Ganghua Chen is an associate professor in the School of Tourism Management, Sun Yat-sen University, China.

Angzhi Gao is an associate professor at the Tourism College of Zhejiang, China.

Huimin Gu is a professor in the School of Tourism Sciences at Beijing International Studies University, China.

Pingping Hou is a postdoctoral fellow in the China Tourism Academy, China.

Jian Hu is the Deputy Director of Cooperation and Development Division of Tourism College of Zhejiang, China.

Songshan (Sam) Huang is a professor in the School of Business and Law at Edith Cowan University, Australia.

Bin Li is an associate professor in the School of Tourism Sciences at Beijing International Studies University, China.

Liang Liu is a PhD candidate in the School of Business and Tourism Management at Yunnan University, China.

Lili Liu is the General Secretary of China Tourism Education Association, China.

Qiuju Luo is a professor in the School of Tourism Management at Sun Yat-sen University, China.

Xianrong Luo is an associate professor in the School of Management at Guangdong Polytechnic Normal University, China.

Xiangru Qin is a PhD candidate in the Research School of Management, the Australian National University, Australia.

Hanjun Tao is a former officer at the China National Tourism Administration, China.

Li Tian is a professor in the School of Business and Tourism Management at Yunnan University, China.

Kunxin Wang is a professor at the Tourism College of Zhejiang, China.

Honggang Xu is a professor in the School of Tourism Management at Sun Yat-sen University, China.

Zhichao Yang is a lecturer at the Tourism College of Zhejiang, China.

Yanbo Yao is a professor in the College of Tourism and Service Management at Nankai University, China.

Xueting Zhai is a PhD candidate in the School of Tourism Management at Sun Yat-sen University, China.

Chaozhi Zhang is a professor in the School of Tourism Management at Sun Yat-sen University, China.

Qingfang Zhang is a PhD candidate in the School of Tourism Management at Sun Yat-sen University, China.

Xiaofeng Zhou is a PhD candidate in the School of Tourism Management at Sun Yat-sen University, China.

Preface

It appears to be challenging to edit such a book as this volume with most of the contributing authors based in China. The book is intended to introduce historical development and the current state of hospitality and tourism education in China to the international audience. To this purpose, it is important to transform some popular terms in China's language system denoting different forms of education, teaching, and training requirements into easy-to-understand words in English so that readers who do not possess due background knowledge of China's education system can understand. Only in the editing process did we realise that language and the Chinese terminology in education do impose a significant level of difficulty in the editing work. Authors usually think about the issues following the Chinese terminology and prepare the chapters based on references in Chinese. Direct translations did not suffice to make the texts understandable. Despite the editors' efforts, there may still be some parts in some chapters that readers may not fully grasp the meaning of the expressions. In this regard, we the editors bear the blame and would like to seek readers' understanding. Chapter authors have been very cooperative to address our critical editing comments and they can only be thanked for their contributions to realise this book!

We give our special thanks to Miss Yuxian Juan, a postdoctoral fellow in the School of Tourism Management at Sun Yat-sen University, for her assistance in editing this book. This book project has been supported by the National Social Science Fund of China key grant (19ZD26) "Strategies, paths and solutions of talent culturation in culture and tourism industries" to Professor Jigang Bao. We also acknowledge the Vice Chancellor's Professorial Research Fellowship provided to Professor Sam Huang by the Edith Cowan University in enabling this book.

We are indebted to our families for their support and understanding to our professional work. Some of the work in this book was completed in the difficult times of lockdown. Our families have granted us their love, patience, and understanding in these unusual times.

Jigang Bao
Songshan (Sam) Huang
August 2021

1 Hospitality and tourism education in China

An overview

Jigang Bao and Songshan (Sam) Huang

Introduction

The scale of the global tourism industry suffices to justify the importance of higher education in hospitality and tourism. According to World Travel and Tourism Council (WTTC, n.d.), in 2019, the travel and tourism sector created 10.4% of the global GDP with 334 million jobs. Every 1 in 10 jobs were in the travel and tourism sector in 2019, and during 2014–2019, a quarter of the new jobs in the world were created in the travel and tourism sector (WTTC, n.d.). The last 40 or so years have witnessed a significant growth of the hospitality and tourism education globally (Airey, 2015; Airey, Tribe, Benckendorff, & Xiao, 2015), corresponding to the remarkable growth of the relevant industry sectors.

In the tourism field, academic attention has been paid to the development and issues of hospitality and tourism education worldwide. Three dedicated journals, namely *Journal of Hospitality and Tourism Education, Journal of Teaching in Travel and Tourism*, and *Journal of hospitality, Leisure, Sport and Tourism Education*, have been in place for more than two decades to publish research in relation to hospitality and tourism education. According to Hsu, Xiao, and Chen (2017), these three journals, together with 10 other mainstream hospitality and tourism journals, published a total of 644 articles pertaining to various issues of hospitality and tourism education in the period 2005–2014. Overall, topics under research covered issues of teaching and learning, student development, curriculum and program, education environment, and faculty development. A number of edited volumes, including Airey and Tribe (2005), Benckendorff and Zehrer (2017), Dredge, Airey, and Gross (2015), Hsu (2005), Liu and Schänzel (2019), and Sheldon and Hsu (2015) added to intensive discussions and debates on the issues of hospitality and tourism education worldwide.

Internationally, hospitality and tourism education is believed to have appeared in some European countries in the 1960s while earlier tourism programs could be traced to the University of Rome dating from 1925 (Airey, 2019; Medlik, 1965). With the growth of tourism industry, tourism education experienced phenomenal growth as well, especially after the turn of the century (Airey et al., 2015).

DOI: 10.4324/9781003004363-1

At the global level, it is worthwhile to recapitulate some of the macro-issues discussed in the literature. First of all, hospitality and tourism education has been operating in a neoliberal institutional environment characterised by managerialism, performativity, market competition, quality assurance and accreditation, and increasing metrification (Ayikoru, Tribe, & Airey, 2009) and privatisation (Hobson, 2010). Second, although hospitality and tourism education started with the vocational needs from the industry sector, in the curriculum space, especially that in higher education, the literature is fraught with debates and arguments to balance a vocational pedagogy with a liberal arts pedagogy (c.f., Dredge, Benckendorff, Day, Gross, Walo, Weeks, & Whitelaw, 2012; Sheldon, Fesenmaier, & Tribe, 2011; Tribe, 2002). Along this stream, of particular note is the Tourism Education Futures Initiative (TEFI), which advocates five clusters of values in tourism education, including stewardship, ethics, knowledge, mutuality, and professionalism (Sheldon et al., 2011; Sheldon, Fesenmaier, Woeber, Cooper, & Antonioli, 2008). Third, internationalisation of tourism education is one of the megatrends (Hobson, 2010; Hsu, 2017). This is especially evident in the higher education system in Western developed countries. For instance, in Australia, universities offering programs in hospitality and tourism rely heavily on international students for the sustaining development and survival of these programs.

Since the turn of the century, China has emerged as one of the most important tourist-origin and destination countries. In 2019, China contributed USD 255 billion or one-fifth of international tourism spending as the top tourism spending country in the world. At the same time, China was ranked No. 4 worldwide receiving 66 million tourist arrivals in 2019 (UNWTO, 2021). With the demand of labour and talents from the rapid industry development, tourism education in China has experienced substantial development. In 2016, there were 1,690 higher education institutions and 924 vocational schools offering hospitality and tourism programs in China. These institutions and schools enrolled about 440,405 tertiary-level students and 232,029 vocational school students, respectively. Such a scale of hospitality and tourism education makes China a large hospitality and tourism education provider in the world.

Despite the scale of China's tourism industry and its tourism education system, studies of hospitality and tourism education in China in the English academic literature (e.g., Gu, Kavanaugh, & Cong, 2007; Gu & Hobson, 2008; Lam & Xiao, 2000; Li & Li, 2013; Wen, Li, & Kwon, 2019; Xiao, 2000; Yin & Meng, 2018; Zhang & Fan, 2005; Zhang, Lam, & Bauer, 2001; Zins & Jang, 2019) seem to have only presented a fragmented and incomplete understanding of China's rather unique and complicated hospitality and tourism education system. As China has its own social and political institutions which define the education system, the hospitality and tourism education system in China demonstrates distinctive features and historical development tracks. There is thus a need to examine China's hospitality and tourism education as a whole in a systemic and holistic way. Such an examination is believed to contribute to the international tourism education scholarship. On one hand, China is still a significant international student market for international tourism education. Understanding the hospitality and tourism education provisions inside China can give international education providers better contextual knowledge to understand Chinese students' learning needs and cultural idiosyncrasies against China's education background. On the other hand, it is likely China's hospitality and tourism education will become more internationalised and international education providers will need to seek more innovative collaboration models with their Chinese counterparts. A systemic examination and understanding of China's hospitality and tourism education can prepare all stakeholders to make better informed strategic moves in the future, given the uncertainties created by the current COVID-19 pandemic.

As such, the current book is intended to give a more holistic and systemic examination of China's hospitality education system. Chapters in this volume present facts about the hospitality and tourism educations in China and provide discussions on the problems and issues facing different subsectors in this education system.

This chapter serves as the introductory chapter for the whole book. In this chapter, we provide an overview of China's hospitality and tourism education. The next section reviews the historical development the hospitality and tourism education in China. We then discuss the current state of China's hospitality and tourism education followed by discussions of some development problems and issues before we conclude this chapter.

Historical development of China's hospitality and tourism education

Substantial development to a modern tourism industry in China started in 1978, when China was in a start-up stage for its reform and opening up. By the time, a pressing problem in China's national economy is the shortage of foreign exchange. Tourism is regarded as a means that can bring the much needed foreign exchange in the country's economic development (Xiao, 2006). To meet the need of earning foreign exchange in the nation's economic development, inbound tourism was prioritised in the early stage of tourism development. With the industry development came the need of personnel and service staff in the industry; thus China's hospitality and tourism education started to grow corresponding to the practical industry needs for qualified labour and workers. In 1979, Shanghai Tourism College was established, marking the beginning of China's higher education in hospitality and tourism (Lam & Xiao, 2000). Since then, a few milestones were realised in the development of China's tourism education. In 1990, the first master's degree program in tourism economics was established in Zhejiang University. In 2000, The School of Management of Sun Yat-sen University started the first PhD program in tourism management, marking the beginning of China's doctoral education in tourism. In 2010, the Academic Degree Committee of the State Council approved the set-up of Master of Tourism Management (MTA) programs in China's university system (Bao, Zhu, & Xin, 2016; Guan & Zhang, 2018).

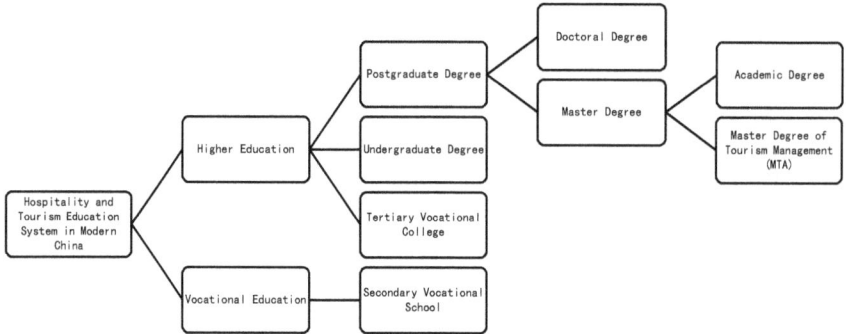

Figure 1.1 The tourism education system of China.

Over the past 40 years, China's hospitality and tourism education has developed to form a complete system which includes four levels of secondary vocational school, tertiary vocational college, undergraduate, and postgraduate degrees, based on the progressive stages of formal education (Figure 1.1). In retrospect, by the end of the 1970s, tourism was still nascent in China; there were very limited tourism personnel and research outputs. Meanwhile, there were prevailing debates and doubts whether tourism can be a subject area of study and a field of scholarship in the society and academia. These cause a certain level of difficulty for hospitality and tourism to be featured and developed in China's higher education system.

Development of undergraduate and postgraduate education

Early tourism higher learning institutions followed three types of governance models: those governed by local/provincial governments and authorities, those dually supported by both the central government bodies and local/provincial authorities, and those directly managed by central government bodies. In the central government system, the first agency in charge of tourism was the Bureau of Travel and Tourism (Zhang, 2003), established in 1964 and affiliated to the Ministry of Foreign Affairs. Later, the Bureau of Travel and Tourism was renamed to China National Tourism Administration (CNTA) in 1982 as a direct affiliate to the State Council. In 2018, CNTA was merged into the Ministry of Culture and formed the currently operating Ministry of Culture and Tourism. In the historic development of China's tourism education, CNTA played a significant role.

The first type of tourism higher learning institutes was supported by local/provincial governments or authorities. The earliest university offering undergraduate degree education in hospitality and tourism is the branch college of Beijing Second Foreign Language Institute founded in 1978, on which basis Beijing Tourism College was established in September 1980 (Zhang,

2018). Beijing Tourism College was affiliated to Beijing Municipal Bureau for travel and tourism (renamed as Beijing Tourism Development Commission in 2011). In April 1985, Beijing Tourism College was incorporated into Beijing Union University and became the Beijing Union University Tourism College. Accordingly, its direct overseeing government body changed to Beijing Municipal Education Commission.

The second type of tourism higher learning institutes was jointly supported by both CNTA and local/provincial universities. Such institutions included Hangzhou University, Nankai University, and Northwest University. In 1980, Hangzhou University established the first undergraduate degree program in tourism economics. Later in 1984, Hangzhou University started to recruit students for its master's program in tourism. In 1987, the program was expanded to form the Department of Tourism in the university, which was further developed to be the Tourism College of Hangzhou University in 1993. In 1998, Hangzhou University merged into Zhejiang University and its tourism college was combined to Zhejiang University's School of Management to form the Department of Tourism and Hotel Management (Zhang, 2018). In September 1981, with the RMB 1.2 million investment from CNTA, Northwest University started its first batch enrolment of students in the tourism economics program. In 1985, CNTA further invested RMB 2.4 million to set up the tourism accounting program. In the same year, the Tourism Department was formally founded at Northwest University, comprising three teaching and research units: Tourism Management and Economics, Foreign Languages, and Accounting. In 1981, CNTA signed an agreement with Nankai University, investing RMB 4.4 million to set up the tourism management and tourism English undergraduate programs in the university's Department of History, supporting the programs to recruit 100 students each year from across the country. In April 1982, Nankai University set up the first Department of Tourism in China, affiliated to the Faculty of Economics.

The third type of tourism higher learning institutes are those universities and colleges directly managed by central government agencies (e.g., CNTA and the Overseas Chinese Office of the State Council). In February 1983, the State Council changed the direct overseeing central government body of Beijing Second Foreign Language Institute (later known as Beijing International Studies University [BISU]) from the Ministry of Education to CNTA. Thus, BISU became the only university directly affiliated to CNTA. In 2000, with the central government restructuring and reforming, BISU was transferred to the Beijing Municipal Government as its overseeing authority. In 1983, Huaqiao University, directly managed by the State Council's Overseas Chinese Office, started to build its Department of Tourism. In 1984, the Department started to recruit undergraduate diploma students in its tour guiding and tourism management programs. Two years later, the Department started to enrol undergraduate degree students. The Department of Tourism was elevated to form the College of Tourism in 2004.

Four universities, that is, Hangzhou University, Northwest University, Nankai University, and BISU, received direct funding support and development guidance from CNTA in their undergraduate tourism programs (Zhang, 2018). Subsequently, in 1986, CNTA continued to cooperate with Sun Yat-sen University to establish the four-year tourist hotel management undergraduate degree program. Enrolment started in 1987 with 20 students and further expanded to 40 students in 1988. Students were recruited across the country from different provinces and were employed through the national university graduate employment system at that time. The tourist hotel management program was later developed into the Department of Hotel Management, hosted in the School of Management at Sun Yat-sen University. In 2004, the School of Tourism Management was founded at Sun Yat-sen University, which later became a leading tourism school in China.

Development of diploma-level tourism and hospitality programs

At the three-year college diploma level, there were three main models of program set up: cooperation between local/provincial government and local/provincial universities; cooperation between CNTA and local/provincial universities; and programs entrusted by tourism enterprises.

Under the first model of cooperation between local government and universities is the establishment of China's first higher learning institute specialising in training high-level tourism personnel, Shanghai Institute of Travel and Tourism (SITT) (上海旅行游览专科学校). SITT was founded in 1979, initially affiliated to the Foreign Affairs Office of Shanghai Municipal Government. In 1980, with the approval of the Ministry of Education, it was renamed Shanghai Tourism Institute (上海旅游专科学校). In 1983, it was dually overseen by the Shanghai Tourism Bureau and the Shanghai Higher Education Bureau, and later in 1986, its overseeing government body was further changed to CNTA. In 1992, the name was changed again to Shanghai Tourism College (上海旅游高等专科学校). In 2000, its affiliation further changed to Shanghai Education Commission. In 2003, Shanghai Tourism College merged into the College of Urban Development and Tourism in Shanghai Normal University to form the current Tourism College at Shanghai Normal University. Another case is Guilin Tourism Institute. Guilin Tourism Institute (桂林旅游专科学校) was first established in November 1985 with the approval of the Government of Guangxi Zhuang Autonomous Region, with the Guilin Municipal Government as its overseeing government body. It started to enrol three-year diploma-level college students in tourism from 1988. In 1994, it was renamed the Guilin Tourism College (桂林旅游高等专科学校), which was further upgraded to Guilin Tourism University (桂林旅游学院) in 2015.

As for the second model, CNTA collaborated with Dalian Foreign Language Institute, Xi'an Foreign Language Institute (renamed as Xi'an Foreign Language University in 2006), Changchun Foreign Language Institute (merged into Changchun University in 1987), respectively, to open joint programs in tourism. In 1979, CNTA cooperated with Dalian Foreign Language Institute to found the Japanese Language Tour Guiding program. Statewide recruitment and enrolment started in 1980. In 1986, CNTA entrusted Xi'an Foreign Language Institute to set up English and Japanese Translation and Tour Guiding Diploma programs in its newly established Department of Tourism (founded in 1985). The study duration was two years (later changed to three years). The programs were open to the whole country for recruitment and graduates are assigned jobs through the national university employment system. In the same year, CNTA entrusted Changchun Foreign Language Institute to set up the two-year diploma-level Japanese Tour Guiding program.

The third model features the cooperation between universities and tourism enterprises. Examples include the hotel management program jointly supported by the Guangdong Pearl River Hotel Management Group and Guangzhou University. In addition, local tourism authorities also entrusted local universities to set up tourism programs. For instance, Hunan Provincial Tourism Administration and Changsha University cooperated in the Department of Tourism of Changsha University; Shandong Provincial Tourism Administration supported Shandong Normal University in establishing its tour guiding and translation diploma programs (Zhang, 2018).

Early scholars and scholarly groups

With the establishment of the relevant tourism and hospitality programs mentioned above, there emerged the early individual tourism scholars/educators and scholarly groups. Prominent early tourism scholars include Professor Guo Laixi in the Institute of Geography of the Chinese Academy of Science (later, in 1999, renamed as the Institute of Geography Science and Resources of the Chinese Academy of Science), Professor Chen Chuankang in Peking University, and some scholars from the Chinese Academy of Social Sciences and Shanghai Academy of Social Sciences (Zhang, 2014).

Development of secondary tourism vocational education

CNTA also played an important role in the beginning and the subsequent development of China's secondary tourism vocational education. In October 1978, China's first vocational tourism school – Jiangsu Province Tourism School – was established in Jiangsu Province. Subsequently, Beijing Tourism Institute, Hubei Tourism School, and Sichuan Tourism School were founded. These four tourism schools constituted the first batch of tourism vocational school in China (Li & Wu, 2018). Later, in the 1980s, secondary tourism vocational schools were founded in other provinces like Tianjin, Shanxi, Liaoning, and Zhejiang. A secondary tourism school usually only had a limited number of programs upon its establishment. For example, when it was founded in 1978, Zhejiang Tourism Vocational School only had two programs: *Hotel Service and Management* and *Travel Agency Service and Tour Guiding*.

In the development process, CNTA provided support to some vocational tourism schools and greatly facilitated the sector's development. The development experiences of early vocational tourism schools provided important references and lessons for the late-coming tourism vocational schools. In 1990, there were already 24 secondary tourism vocational schools, 136 vocational high schools, and ordinary high schools offering tourism programs countrywide.

Current state of hospitality and tourism education in China

After over 40 years development, hospitality and tourism education in China is facing a new landscape. Developments in both quality and quantity have been witnessed in the increasing numbers of tourism education institutions, student enrolments, and teaching faculty.

Scale of institutions and students

Tourism education institutions are the foundation and main actors of the tourism education system. In 2010, the number of students enrolled in the hospitality and tourism education institutions in China broke the threshold of one million for the first time; and the total number of tourism education institutions exceeded 2,000. From 2013 to 2017, the number of commencing student in the hospitality and tourism education institutions each year was between 250,000 and 300,000 (Table 1.1). By 2017, the total number of hospitality and tourism education institutions reached 2,641 (Figure 1.2), and the overall growth rates in the number of hospitality and tourism education institutions from 1993 to 2017 are generally positive (Figure 1.3). In terms of total student enrolment number, in 2016, a total of 672,434 students were enrolled in the hospitality and tourism programs in these institutions in China (Figure 1.4). As shown in Figures 1.4 and 1.5, the growth of China's hospitality and tourism education in terms of student enrolment scale has slowed down in recent years. In 2016, there were 1,690 higher learning institutes (including tertiary vocational colleges and institutes) in China, enrolling a total of 440,404 students; the same year saw a total of 232,029 students enrolled in 924 secondary vocational schools in China.

Table 1.1 Recruitment numbers of China's tourism education institutions (2013–2017) (unit: person)

Year	Doctorate	Masters	Undergraduate	Tertiary diploma	Secondary vocational	Total
• 2013	• 200	• 1,600	• 52,100	• 129,100	• 117,000	• 300,000
• 2014	• 167	• 1,569	• 53,386	• 110,835	• 123,000	• 288,957
• 2015	• 257	• 1,619	• 55,611	• 110,935	• 93,000	• 261,422
• 2016	• 360	• 1,679	• 58,000	• 116,000	• 104,000	• 280,039
• 2017	• 336	• 2,832	• 59,000	• 113,000	• 102,000	• 277,168

Data source: Statistics of national tourism education and training, CNTA.

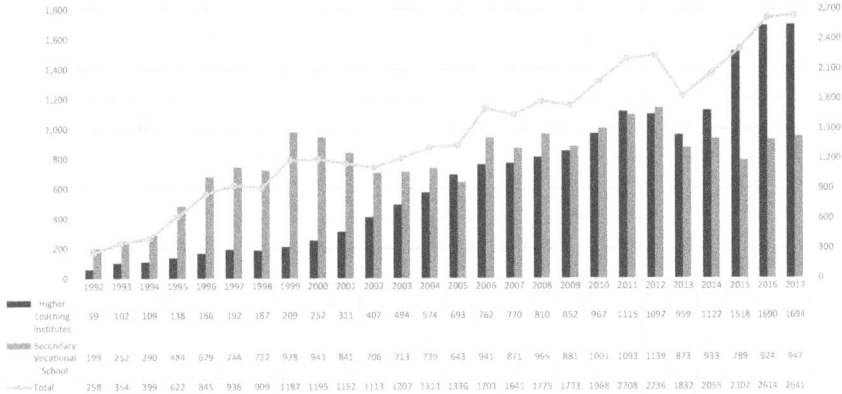

	1992	1993	1994	1995	1996	1997	1998	1999	2000	2001	2002	2003	2004	2005	2006	2007	2008	2009	2010	2011	2012	2013	2014	2015	2016	2017
Higher Learning Institutes	59	102	109	138	186	192	187	209	252	311	407	484	574	693	762	770	810	852	967	1115	1097	959	1122	1518	1690	1684
Secondary Vocational School	199	252	290	484	679	744	722	978	943	841	706	713	739	643	941	871	965	881	1001	1093	1139	873	933	789	924	947
Total	258	354	399	622	845	936	909	1187	1195	1152	1113	1207	1313	1336	1703	1641	1775	1733	1968	2208	2236	1832	2055	2107	2614	2641

Figure 1.2 Number of tourism education institutions in China (1992–2017).
Data source: China Tourism Statistic Yearbooks.

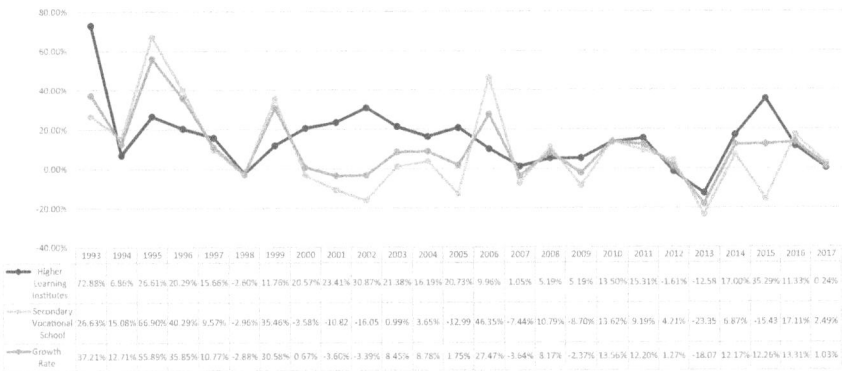

	1993	1994	1995	1996	1997	1998	1999	2000	2001	2002	2003	2004	2005	2006	2007	2008	2009	2010	2011	2012	2013	2014	2015	2016	2017
Higher Learning Institutes	72.88%	6.86%	26.61%	20.29%	15.66%	-2.60%	11.78%	20.57%	23.41%	30.87%	21.38%	16.19%	20.73%	9.96%	1.05%	5.19%	5.19%	13.50%	15.31%	-1.61%	-12.58	17.02%	35.29%	11.33%	0.24%
Secondary Vocational School	26.63%	15.08%	66.90%	40.29%	9.57%	-2.96%	35.46%	-3.58%	-10.82	-16.05	0.99%	3.65%	-12.99	46.35%	-7.44%	10.79%	-8.70%	13.62%	9.19%	4.21%	-23.35	6.87%	-15.43	17.11%	2.49%
Growth Rate	37.21%	12.71%	55.89%	35.85%	10.77%	-2.88%	30.58%	0.67%	-3.60%	-3.39%	8.45%	8.78%	1.75%	27.47%	-3.64%	8.17%	-2.37%	13.56%	12.20%	1.27%	-18.07	12.17%	12.26%	13.31%	1.03%

Figure 1.3 Growth rates of the number of tourism education institutions in China (1993–2017).
Data source: China Tourism Statistic Yearbooks.

It should be noted that before 2005, the number of secondary vocational schools and their enrolled students was much more than those of higher learning institutes; however, in 2005, the number of tertiary institutes and their enrolled students for the first time exceeded those of secondary vocational schools. Until 2011, the growth rates of student enrolments in tertiary institutes were much higher than that of second vocational schools. From 2013 on, the increase of tertiary institutes was apparent, far exceeding the growth of secondary vocational schools. China's tourism education demonstrated a phenomenon of "hot higher education and cool secondary vocational education". According to CNTA, the number of students applying for second tourism vocational schools

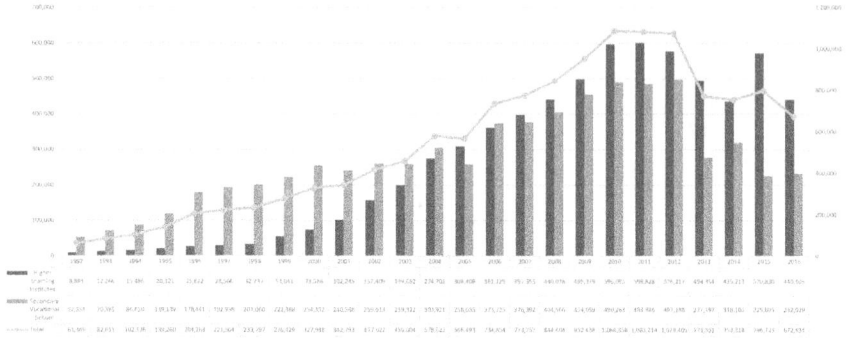

Figure 1.4 Enrolment numbers of tourism education institutions in China (1992–2016).
Data source: China Tourism Statistic Yearbooks.

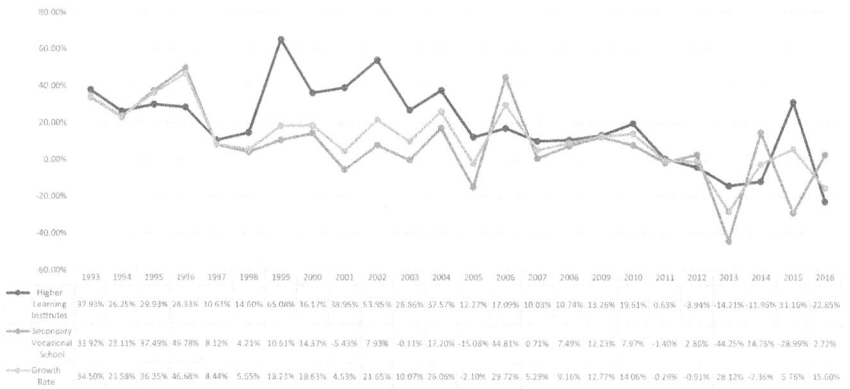

Figure 1.5 Growth rates of student enrolment number (1993–2016).
Data source: China Tourism Statistic Yearbooks.

was dwindling, and applications from urban families were especially rare. There
was an obvious declining secondary tourism vocational education, prompting a
critical issue in the realm for tourism educators in China.

Teaching staff

From 2010 to 2012, the total number of teaching staff in China's tourism education system was about 45,000. Fifty-five per cent of these teaching staff are
employed by tertiary education institutes, with an average of 24 staff per institute;
45% work in secondary tourism vocational schools, with an average number of
20 per school (Table 1.2). According to incomplete statistics from CNTA, there
were over 10,000 full-time teaching staff with senior titles (Associate Professor

Table 1.2 Number of teaching staff in China's tourism education system (2010–2012) (unit: 1,000 persons)

	Tertiary institute	Percentage	Secondary school	Percentage	Total
• 2010	• 24.0	• 55.63	• 20.0	• 44.37	• 44.0
• 2011	• 25.3	• 56.22	• 19.7	• 43.78	• 45.0
• 2012	• 24.7	• 54.55	• 19.7	• 45.45	• 4.4.4

Data source: Statistics of national tourism education and training, CNTA.

level or above in universities and senior lecturer or above in secondary schools). Among them, nearly 8,000 were in tertiary institutes and about 2,700 worked in secondary vocational schools. Relative to the large numbers of students in tourism education institutions, the number of full-time teaching staff is still low. There is a pressing need to further develop the teaching staff force in China's tourism education.

Geographical distribution of institutions

The development pace of China's tourism education has been very fast. Right now, all 31 provinces/municipalities/autonomous regions in China have universities/schools offering hospitality and tourism programs. In general, the geographical distribution of tourism education institutions corresponds to the regional economic development levels and tourism development levels. There are more tourism education institutions in tourism-developed regions (Table 1.3). Overall, the difference between the number of tourism education institutions in eastern coastal regions and that in western inland regions is pronounced.

From a regional perspective, until 2017, the East China region hosted the largest number of tourism education institutions, reaching 700, roughly 3.7 times of the least number in the Northeast region. The Southwest region with 571 institutions ranked second. The Central China, South China, and North China regions ranked from third to fifth with 349, 310, and 293 institutions, respectively. The Northwest region, ranking sixth, had 230 institutions. The regions hosting more tourism education institutions, East China, Central China, South China, and North China, are also more developed regions economically and in terms of tourism. The Southwest region possesses abundant tourism resources; accordingly, its number of secondary vocational schools is big, raising up the scale of the whole region's tourism education. Comparatively, Northeast and Northwest had low levels of tourism education corresponding to their low level of economic development.

Examining the numbers in the provinces, it was found that Beijing, Jiangsu, Anhui, Hunan, Guangdong, Sichuan, Chongqing, and Yunnan each had over 100 tourism education institutions. In contrast, Tibet, Ningxia, Qinghai, Inner Mongolia, Gansu, and Guizhou had less than 50 tourism education institutions each. In terms of tourism education, there is still a big gap between developed provinces and underdeveloped provinces/regions.

Geographic distribution of teachers

According to CNTA, in 2010, significant variations existed among China's tourism education institutions in terms of the ratio of senior full-time teaching staff to total full-time teaching staff. Shaanxi, Qinghai, and Beijing had the ratio above 40%; those provinces possessing a ratio of between 30% and 40% included Hubei, Hebei, Hunan, and Shandong; Jiangxi, Guizhou, Hainan, Henan, Guangxi, and Fujian had a ratio of between 20% and 30%, while Tianjin, Gansu, and Shanxi had a ratio of below 5%.

The student-teacher ratio refers to the ratio between the number of enrolled students and that of registered teachers in the workplace. It is one of the important indicators to measure the quality and effect of education. There are also variations in student-teacher ratios in tourism education in different parts of China. Tables 1.3 and 1.4 show the clusters of provinces in the different categories of student-teacher ratio in their tertiary and secondary tourism education systems, respectively.

In the *Basic Infrastructure Indicators of Higher Education Provisions (Trial)* issued by the Ministry of Education in 2004, the required student-teacher ratio in comprehensive higher leaning institutes is 18:1. Apparently, in the tourism education sector, most provinces had a much higher student-teacher ratio than requested by the Ministry of Education. The tourism education sector in China still faces shortage of qualified teaching staff, with most of the current teachers facing high workload and workplace pressure.

Table 1.3 Student-teacher ratios in tertiary institutions (2010–2011)

Year	Ratio range			
	• >40:1	• 30:1–40:1	*20:1–30:1*	*<20:1*
2010	Hainan, Tianjin, Ningxia, Anhui	Heilongjiang, Gansu, Hunan, Shaanxi	Hebei, Shanxi, Yunnan, Inner Mongolia, Tibet, Liaoning, Shanghai, Jiangsu, Zhejiang, Jiangxi, Shandong, Henan, Guangdong, Guangxi	Guizhou, Xinjiang, Hubei, Fujian, Sichuan, Qinghai, Chongqing, Beijing, Jilin
2011	Tianjin, Jiangsu, Beijing, Anhui	Hainan, Henan, Liaoning, Hunan, Shaanxi, Ningxia	Hebei, Shanxi, Inner Mongolia, Jilin, Heilongjiang, Shanghai, Zhejiang, Fujian, Jiangxi, Shandong, Hubei, Guangdong, Guangxi, Sichuan, Chongqing, Guizhou, Yunnan, Tibet, Gansu, Qinghai, Xinjiang	

Data source: Statistics of national tourism education and training, CNTA.

Table 1.4 Student-teacher ratios in secondary tourism vocational schools (2010–2011)

Year	Ratio range			
	• >40:1	• 30:1–40:1	*20:1–30:1*	*<20:1*
2010	Zhejiang, Chongqing, Tianjin	Jiangsu, Hainan, Beijing	Inner Mongolia, Shanghai, Fujian, Shandong, Hubei, Hunan, Guangdong, Guangxi, Yunnan, Tibet, Shaanxi, Gansu, Ningxia	Jiangxi, Sichuan, Guizhou, Anhui, Henan, Shanxi, Heilongjiang, Liaoning, Xinjiang, Qinghai, Hebei, Jilin
2011	Hainan, Guangdong, Tianjin, Chongqing	Shandong, Jiangsu, Gansu, Yunnan	Beijing, Hebei, Shanxi, Inner Mongolia, Liaoning, Jilin, Heilongjiang, Shanghai, Zhejiang, Anhui, Fujian, Jiangxi, Henan, Hubei, Hunan, Guangxi, Sichuan, Guizhou, Tibet, Shaanxi, Qinghai, Ningxia, Xinjiang	

Data source: Statistics of national tourism education and training, CNTA.

Problems and challenges

Tourism has developed to be an important industry in China and has become a significant component in the national strategy. However, tourism education, as an indispensable system to produce qualified workers and talent to the industry, is still facing a number of serious problems and challenges. The existence and development of tourism education relies on the tourism industry development. Tourism industry development needs to promote the growth of tourism education in any country. There is a natural symbiosis relationship between tourism education and tourism industry development. Currently, the problems and challenges facing China's tourism education are mainly reflected in the following five aspects: the structure of tourism education system, teaching faculty development, curriculum and textbook standard, first-level subject construction, and practice teaching.

Tourism education structure needs to be optimised

Although the enrolment scale of China's tourism education has been evergrowing, the structure of tourism education needs to be optimised in order to meet the needs of the tourism industry and the society (Ji & Lu, 2005; Zhang, 2016). In 2017, the number of commencing students at postgraduate (including doctoral, masters, and professional master's students), undergraduate, tertiary vocational diploma level, and secondary vocational school level in China's tourism education was 4,270, 59,000, 113,000, and 102,000, respectively, exhibiting a pyramid shape with a contracting base. Geographically, the level of tourism education in eastern coastal provinces is significantly higher than that in western inland areas.

While there is an oversupply of tertiary-level undergraduate degree graduates, the number of graduates from secondary vocational schools is relatively low. A lot of tourism enterprises in the industry cannot find satisfactory workers from the pool of tourism major graduates; tourism students also cannot find jobs to their satisfaction. There is a structural disequilibrium between the supply and demand of tourism industry workers. The basic reasons may be in the tourism education modes and its structure. Tourism is an applied subject. Tourism education requires the active involvement and coordination of multiple stakeholders, including government, industry, and academia (Zhang, 2016). Tourism education in China should further optimise the structure of different levels, so that it can meet the demand of the tourism industry. The future development of tourism education should match the workforce needs imposed by the industry development. In this regard, it is important to consider the different layers or levels of personnel, i.e., research-oriented, management practice-oriented, and service-oriented, to suit the development needs of different organisations such as research institutions, government agencies, and various types of tourism enterprises.

The quality of tourism teach staff has much variation

In the process of tourism education, student learning outcome is largely dependent on the teaching process, which is mostly determined by the teaching staff based on the teaching materials. Therefore, teaching staff and textbooks are important elements in the teaching and learning process (Bao, Xie, Wang, Ma, & Xiao, 2019). The main problem with teaching staff in tourism education is the varying quality of the teaching workforce. There are three underlying issues. First, there are wide differences in the disciplinary backgrounds the hospitality and tourism programs across different institutions. Most of the programs have the disciplinary backgrounds of geography, economics, foreign languages, and history. Tourism subject teachers also have different disciplinary backgrounds. In most cases, courses are set up based on the disciplinary expertise of the teachers, rather than what needs to be taught in the program (Bao et al., 2019; Wu & Ye, 2004). Most of the teaching staff entered the field of tourism education from other disciplines and are facing the needs to adapt themselves to the new teaching field. They also need to develop their research and teaching capability in their career. If they cannot conceptualise and theorise the newly emerging tourism industry phenomena in their teaching process, they will not be able to conduct the teaching to prepare students to cope with the fast-changing industry realities (Bao et al., 2019). Second, under the current performance assessment system in China's higher education sector, pragmatism prevails among teaching staff. High pressure on university staff for research outputs makes little time for them to concentrate on the quality of teaching. Some teaching staff are over-reliant on the traditional approach of teaching methods and the outdated textbooks and knowledge system, and have little motivation to update the curricula and teaching contents and engage in teaching innovations. As a result, little emphasis has been laid on students' critical thinking and problem-solving abilities. Finally, tourism education should complement the development of tourism industry. This requires both teachers and students in the education process to possess both knowledge of the field and the practical industry experience and capabilities. However, in the current tourism education system, most of the teaching staff in hospitality and tourism programs do not have prior industry experience, making them less capable to adapt the new industry developments into their teaching and prepare students to be job-ready (Zhao, 2020).

Tourism textbooks and curricula have differing standards

Textbooks and teaching contents are the core instrument in the teaching and learning process. In terms of textbooks, there are two problems. First, some publishing houses are too profit-driven; they publish some "series textbooks" based on the market needs of the tourism education stakeholders in order to maximise the market share. Second, some education institutions regard textbooks publication as a performance indicator for staff promotion, thus causing tourism academics herd themselves in quick and low-quality textbook writing and publishing (Xie, 2008). There is a need to balance the quality and quantity of tourism textbooks.

In the curriculum space, a lot of debates have been around the designation of the core curriculum in hospitality and tourism programs. There are three reasons: one, different institutions have their tourism programs in different disciplinary backgrounds; there has been a strong path on dependence in the curricula and little consensus can be achieved. Two, different levels of education offerings have different positioning for their programs and target student markets. While some universities focus more on teaching theories, other institutions may emphasise on vocational skills in their teaching; accordingly, there are apparent differences in the designations of curricula. Three, currently tourism-related programs include three-degree tracks: tourism management, hotel management, and event economics and management. These three majors differ a lot in their teaching practices and research, causing different opinions in setting the core courses in the programs.

The status of tourism subject in the education system is low

To move the tourism programs' curriculum and textbook development into further maturity, consensus should be built on the core knowledge of tourism programs in the academia (Bao et al., 2019). However, there still seems to be a long way to get the tourism subject to be recognised as a first-level academic subject in China's education system. In the undergraduate degree programs catalogue promulgated by the Ministry of Education in 2014, tourism management was promoted to be a first-level subject. However, in the State Council Degree Office's Degree Awards Catalogue and the Academic Subjects Catalogue of Postgraduate Degree Programs, tourism management still remains to be a second-level subject. Such a configuration imposes significant constraints for China's tourism education development. As resource allocation in universities is mainly based on the first-level subjects, some tourism colleges cannot obtain the eligibility to recruit postgraduate students. Some hospitality and tourism programs face the fate of reduction, merging into other programs, and even discontinuation. Tourism schools and programs also face difficulty to recruit teaching staff from other disciplines.

There have been increasing discussions on the discipline nature of tourism internationally. From an applied perspective, the industries and sectors corresponding to tourism and tourism activities are clear; also, in the history of tourism education development, tourism scholars have been attempting to build their own core concepts and theories/theoretical frameworks. A common theoretical and epistemological foundation has been in formation in the tourism academia (Bao, 2015). For the time being, tourism has demonstrated a lot of characteristics of a first-level subject. Therefore, further promoting the status of the tourism subject and elevating tourism as a first-level subject in the education system can effectively resolve the problems and further raise the quality of tourism education in China.

Practice-based teaching needs to be promoted and strengthened

Practice-based teaching and learning seems to be an indispensable part of tourism education. Current modes of practice-based teaching mainly include field trips and observation, scenario-based teaching, and internship (Han & Lu, 2010). Around practice-based teaching, there are three main problems. First, in many universities, the practice-based teaching mode is rigid and inflexible, lacking stable on-campus and off-campus practice base; students cannot master the knowledge and skills needed by the industry (Han & Lu, 2010). Second, in the practices of university-enterprise cooperation, enterprises lack interest in students and the goals of universities do not match the interest of enterprises; students are not satisfied with enterprises (Zhao & Tang, 2005). Third, the ideology of balancing liberal arts with vocational skills need to be further reinforced. In China's tourism education ideology, vocational courses aim to prepare applied technical and skill workers, while higher education courses are intended to produce knowledge type and comprehensive personnel. Either side is more or less biased to foster one single type of quality or skills, thus overlooking the balance between liberal arts and vocational education. In summary, there are many practical problems around the practice-based teaching in China's tourism education system. Resolving the problems would involve concerted efforts among universities, enterprises, and students, and ideological changes.

Conclusion

This chapter serves as the introductory chapter to the whole book. It opens with a grand contextual discussion of hospitality and tourism education internationally.

Following it is a brief review of the historical development of China's hospitality and tourism education. Some early development facts are provided. The chapter further goes for a review of the current state of hospitality and tourism education, mainly in the aspects of scales of the institutions and students, teaching staff, geographical distributions of institutions, students, and teaching staff. While readers can expect to develop a better understanding of the development problems and issues in the specific sub-areas of China's hospitality and tourism education in the following chapters, this chapter briefs on the commonly agreed general problems and challenges, which include the structural imbalance of tourism education provisions, problems with the teaching staff quality, low quality of textbooks and unstandardised curriculum articulations, low status of tourism subject in the institutional system, and problems with practice-based teaching.

References

Airey, D. (2015). 40 years of tourism studies – A remarkable story. *Tourism Recreation Research, 40,* 6–15.

Airey, D. (2019). Education for tourism: A perspective article. *Tourism Review,* DOI: 10.1108/TR-02-2019-0074.

Airey, D., & Tribe, J. (Eds.). (2005). *An international handbook of tourism education.* New York: Routledge.

Airey, D., Tribe, J., Benckendorff, P., & Xiao, H. (2015). The managerial gaze: The long tail of tourism education and research. *Journal of Travel Research, 54*(2), 139–151.

Ayikoru, M., Tribe, J., & Airey, D. (2009). Reading tourism education: Neoliberalism unveiled. *Annals of Tourism Research, 36*(2), 191–221.

Bao, J. (2015). Building tourism management into a first-level subject and accelerating tourism education development. *Tourism Tribune, 30*(9), 1–2.

Bao, J., Xie, Y., Wang, N., Ma, B., & Xiao, H. (2019). Conversations among five tourism scholars: Forty years of tourism education—confusions in no-confusion age. *Tourism Forum, 12*(2), 1–13.

Bao, J., Zhu, L., & Xin, X. (2016), Report on China's tourism postgraduate education (2014-2015). In China Tourism Education Association (Ed.) *Tourism education bluebook of China* (pp. 24–44). Beijing: China Tourism Press.

Benckendorff, P., & Zehrer, A. (Eds.). (2017). *Handbook of teaching and learning in tourism.* Cheltenham: Edward Elgar.

Dredge, D., Airey, D., & Gross, M.J. (Eds.). (2015). *The Routledge handbook of tourism and hospitality education.* New York: Routledge.

Dredge, D., Benckendorff, P., Day, M., Gross, M.J., Walo, M., Weeks, P., & Whitelaw, P. (2012). The philosophic practitioner and the curriculum space. *Annals of Tourism Research, 39*(4), 2154–2176.

Gu, H., & Hobson, P. (2008). The dragon is roaring... The development of tourism, hospitality & event management education in China. *Journal of Hospitality & Tourism Education, 20*(1), 20–29.

Gu, H., Kavanaugh, R.R., & Cong, Y. (2007). Empirical studies of tourism education in China. *Journal of Teaching in Travel & Tourism, 7*(1), 3–24.

Guan, J., & Zhang, C. (2018). Report on China's tourism postgraduate education (2016-2017). In China Tourism Education Association (Ed.) *Tourism education bluebook of China* (pp. 18–28). Beijing: China Tourism Press.

Han, B., & Lu, P. (2010). A Sino-foreign comparison practice-based teaching in tourism higher education. *Human Geography, 25*(6), 154–157.

Hobson, J.S.P. (2010). Ten trends impacting international hospitality and tourism education. *Journal of Hospitality and Tourism Education, 22*(1), 4–7.

Hsu, C.H.C. (Ed). (2005). *Global tourism higher education: past, present, and future.* New York: Routledge.

Hsu, C.H.C. (2017). Internationalisation of tourism education. In P. Benckendorff, & A. Zehrer (Eds.) *Handbook of teaching and learning in tourism* (pp. 321–335). Cheltenham: Edward Elgar.

Hsu, C.H.C., Xiao, H., & Chen, N. (2017). Hospitality and tourism education research from 2005 to 2014: "Is the past a prologue to the future"? *International Journal of Contemporary Hospitality Management, 29*(1), 141–160.

Ji, P., & Lu, J. (2005). Multi-layer tourism education model based on the human resource demand. *Tourism Tribune, 20*(S1), 57–61.

Lam, T., & Xiao, H. (2000). Challenges and constraints of hospitality and tourism education in China. *International Journal of Contemporary Hospitality Management, 12*(5), 291–295.

Li, L., & Li, J. (2013). Hospitality education in China: A student career-oriented perspective. *Journal of Hospitality, Leisure, Sport & Tourism Education, 12*, 109–117.

Li, C., & Wu, S. (2018). An outline of China's tourism vocational education development in recent 40 years and some thoughts. In J. Bao (Ed.) *Tourism education in China: Development, programs and challenges* (pp. 97–135). Beijing: China Tourism Press.

Liu, C., & Schänzel, H. (Eds). (2019). *Tourism education and Asia.* Singapore: Springer Nature Singapore.

Medlik, S. (1965). *Higher education and research in tourism in Western Europe.* London: University of Surrey.

Sheldon, P.J., Fesenmaier, D.R., Tribe, J. (2011). The Tourism Education Futures Initiative (TEFI): Activating change in tourism education. *Journal of Teaching in Travel & Tourism, 11*(1), 2–23.

Sheldon, P., Fesenmaier, D., Woeber, K., Cooper, C., & Antonioli, M. (2008). Tourism Education Futures, 2010-2030: Building the capacity to lead. *Journal of Teaching in Travel and Tourism, 7*(3), 61–68.

Sheldon, P., & Hsu, C. (Eds). (2015). *Tourism education: Global issues and trends.* Bingley: Emerald.

Tribe, J. (2002). The philosophic practitioner. *Annals of Tourism Research, 29*(2), 228–257.

UNWTO. (2021). International Tourism Highlights 2020 Edition. Retrieved June 10, 2021 from http://www.e-unwto.org/doi/pdf/10.18111/9789284422456.

Wen, H., Li, X., & Kwon, J. (2019). Undergraduate students' attitudes towards and perceptions of hospitality careers in mainland China. *Journal of Hospitality & Tourism Education, 31*(3), 159–172.

WTTC. (n.d.). Economic impact reports. Retrieved June 9, 2021 from https://wttc.org/Research/Economic-Impact.

Wu, G., & Ye, X. (2004). Tourism personnel training and the vocational development of higher tourism education. *Tourism Tribune, 19*(S1), 15–18.

Xiao, H. (2006). The discourse of power: Deng Xiaoping and tourism development in China. *Tourism Management, 27*(5), 803–814.

Xiao, H. (2000). China's tourism education into the 21st century. *Annals of Tourism Research, 27*(4), 1052–1055.

Xie, Y. (2008). Issues of textbook construction in China's higher tourism education. *Tourism Tribune, 23*(1), 5–6.

Yin, Z., & Meng, F. (2018). Tourism higher education in China: Profile and issues. In J. Zhao (Ed.) *The hospitality and tourism industry in China: New growth, trends, and developments* (pp. 241–261). Palm Bay, FL: Apple Academic Press.

Zhang, G. (2003). China's tourism since 1978: Policies, experiences and lessons learned. In A.A. Lew, L. Yu, J. Ap, & G. Zhang (Eds). *Tourism in China* (pp. 13–340). New York: The Haworth Hospitality Press.

Zhang, L. (2014). Development process and trend of modern China's tourism education. In China Tourism Education Association (Ed.) *Tourism education bluebook of China* (pp. 2–8). Beijing: China Tourism Press.

Zhang, L. (2016). New opportunities and challenges of China's tourism education and tourism subject development. In China Tourism Education Association (Ed.) *Tourism education bluebook of China* (pp. 2–16). Beijing: China Tourism Press.

Zhang, L. (2018). The horn of reform: The origin and first decade development of China's tourism higher education. In China Tourism Education Association (Ed.) *Tourism education bluebook of China* (pp. 2–12). Beijing: China Tourism Press.

Zhang, W., & Fan, X. (2005). Tourism higher education in China. *Journal of Teaching in Travel & Tourism, 5*(1–2), 117–135.

Zhang, H.Q., Lam, T., & Bauer, T. (2001). Analysis of training and education needs of mainland Chinese tourism academics in the twenty-first century. *International Journal of Contemporary Hospitality Management, 13*(6), 274–279.

Zhao, A. (2020). Development strategies for "dual-capability" teaching force in the context of industry-education integration – the example of tourism management program in Liaodong College. *Journal of Liaodong College (Social Science Edition), 22*(1), 111–116.

Zhao, P., & Tang, L. (2005). A study of the university-enterprise cooperation mechanism in tourism higher education institutions – The example of the Beijing Union University Tourism College. *Tourism Tribune, 20*(S1), 71–76.

Zins, A.H., & Jang, S.Y. (2019). Review and assessment of academic tourism and hospitality programmes in China. In C. Liu & H. Schänzel (Eds). *Tourism education and Asia* (pp. 81–105). Singapore: Springer Nature Singapore.

2 Tourism management undergraduate programs

Li Tian and Liang Liu

Introduction

In the past more than 40 years, China's tourism management undergraduate programs went through three stages: inception, growth, and standardization. In this process, tourism management has transitioned from a single program to a program cluster, including tourism management, hospitality management, exhibition economy and management, and tourism management and service education. Along with China's accelerated promotion of the connotative development of higher education, tourism management undergraduate programs are facing new challenges and problems. How to address these challenges and problems has become a common concern in China's tourism education and industry sectors. Based on a review of the development stages of tourism management undergraduate programs, this chapter analyses its development status from five perspectives: program classification, development scale, regional distribution, university structure, and student numbers. In addition, the talent training status is explored by sorting out forms of enrolment, training goals, curriculum system, faculty, and quality management. Finally, the approaches to constructing first-class programs, designing first-class curricula, and cultivating first-class talents are discussed in detail.

Since the reform and opening-up, China's tourism industry has undergone substantial development and China has become an important tourist destination in the world. According to UN World Tourism Organization (UNWTO), China ranked fourth in the number of international tourist arrivals and continues to lead global outbound travel in terms of expenditure in 2018 (UNWTO, 2019). These data fully illustrate the important position of China's tourism industry in the world. With continuous development of China's tourism industry, tourism management undergraduate programs have undergone the inception and growing stages and come into the stage of standardization. The development has shown a strong and dynamic trend, which is demonstrated by, for example, the establishment of the Tourism Management Teaching Guidance Committee (TMTGC), the promulgation of national standards for teaching quality, and the certification of national first-class programs.

For the time being, a few contextual issues need to be considered for the development of tourism management undergraduate programs in China. First,

DOI: 10.4324/9781003004363-2

the successive initiatives of the "Double First-Class Plan," the "Double Ten-Thousand Plan," and the "Six Excellence and One Brilliance Plan 2.0" are aimed to promote the connotative development of higher education through reconstructing the development strategies and forming a new pattern focusing on disciplines, curricula, courses, and talent training. Second, two university development principles, namely "taking undergraduate education as the core" and "improving the ability of colleges and universities to serve economic and social development" in the tertiary education sector, will be exerting their influence on the development of tourism management undergraduate programs. Third, tourism management undergraduate programs pay more attention to the comprehensive quality, practical skills, and innovative ability of students. In such a context, tourism management undergraduate education needs to rely on industry practice and use modern technology to promote the construction of first-class programs, first-class curricula, and first-class talent training. To this end, it is necessary to systematically research and review the development history as well as the current status of tourism management undergraduate programs to identify current problems and challenges and finally propose relevant constructive solutions and ways to move forward.

Historical review

Tourism management undergraduate programs form the foundation for supporting the development of the tourism industry, and have gradually evolved over a period of 40 years to become an extensive programs system, including tourism management, hospitality management, exhibition economy and management, and tourism management and service education. This process started from the establishment of a number of tourism colleges and universities, most notably the Shanghai Institute of Tourism formed in 1979. Ever since, the development of China's tourism higher education has experienced three developmental stages: the inception stage between 1979 and 1998, the growth stage between 1999 and 2012, and the standardizing stage after 2013. These three development stages of China's tourism higher education are mostly in accordance with the development stages of China's tourism industry (Xia & Xu, 2018).

1978–1998: inception stage

In this stage, China's tourism higher education, like the development of the tourism industry, is in the changing environment of the times. The positioning of China's tourism industry began to transform into promoting economic development, where the tourism industry management shifted from basic administrative management to more specialized industry management. The comprehensive standards, service standards, and facilities standards are incorporated into the responsibilities of the China National Tourism Administration (CNTA; now reorganized into the Ministry of Culture and Tourism [MCT]). At the same time, in order to further stimulate market vitality, the strategy of jointly developing inbound tourism, domestic tourism, and outbound tourism was formulated. Accordingly, the demand for skilled and managerial talents by travel companies, restaurants, scenic spots, and other tourism companies started to increase. It is against this background that China's tourism higher education started to take shape.

The beginning of China's tourism higher education was marked by the following events, which are summarized in Table 2.1. In 1979, the establishment of the Shanghai Institute of Tourism opened China's tourism higher education. The school was affiliated to the CNTA (now the MCT) and its main task was to cultivate senior tourism talents to serve the needs of the country's reform and opening-up. In 1981, the State Education Commission of China (now the Ministry of Education, MOE) approved the establishment of undergraduate tourism programs at the following institutions: Nankai University, Beijing International Studies University, Hangzhou University, Dalian University of Foreign Languages, Northwest University, Xi'an International Studies University, Changchun University, and Sun Yat-sen University. This represented the start of undergraduate education in tourism management in China. Subsequently, many colleges and universities set up tourism-related programs based on their own disciplinary advantages.

China's tourism management had also undergone changes in the way it is presented in the national program catalogue. In 1985, the State Education Commission (now the MOE) began to revise the 1963 edition of the *Universal Programs Catalogue in Colleges and Universities* and issued a new edition in 1989. In the new version of *Social Science Undergraduate Programs Catalogue and Introduction in Colleges and Universities*, Tourism Economics (1018) was presented under Economics and Management Science, and Hospitality Enterprise Management (1045) was also set up as a trial program. This revision effectively categorizes the tourism undergraduate program as a subject in its own right for the first time. In 1993, the *Undergraduate Programs Catalogue and Introduction in Colleges and Universities* revised by the State Education Commission (now the MOE) adjusted the tourism undergraduate program. Tourism Economics was withdrawn and Tourism Management (020209) was established as a program under Business Management (0202) within Economics (02). This revision emphasized the management nature of the tourism undergraduate program. In the 1998 edition of the *Undergraduate Programs Catalogue in Colleges and Universities*, the MOE once again adjusted the home discipline of tourism management programs. Specifically, Management Science (11) was separated from the original Economics discipline and became an independent discipline category. Correspondingly, Tourism Management (110206) became a program in Business Management (1102) under Management Science.

1999–2012: growth stage

The growth stage witnessed a period of continuous adjustment and optimization of tourism management undergraduate programs as well as a period of mass development associated with tourism activities. In 1999, the State Council

Table 2.1 Important events in the inception stage of tourism higher education

Year	Events	Implications
1979	Establishment of the Shanghai Institute of Tourism	Opening China's tourism higher education
1981	Eight universities set up tourism undergraduate programs	Opening China's tourism undergraduate education
1989	Revising *Social Science Undergraduate Programs Catalogue and Introduction in Colleges and Universities*	Identifying tourism economics undergraduate program
1993	Revising *Undergraduate Programs Catalogue and Introduction in Colleges and Universities*	Identifying tourism management undergraduate program
1998	Revising *Undergraduate Programs Catalogue in Colleges and Universities*	Adjusting the attribution of disciplines in tourism management

issued the *National Holiday and Memorial Day Measures* and decided to combine the Spring Festival, Labor Day, and National Day with the two days before and after the holiday to form a seven-day golden week system. This policy greatly promoted the development of both domestic tourism and outbound tourism. In 2001, China officially joined the World Trade Organization and made specific commitments in the market access and treatment for foreign company in the tourism industry, which brought both opportunities and challenges to the development of China's inbound and outbound tourism (Zhang & Fan, 2005). In 2009, the State Council released the *Guidelines of Accelerating the Development of Tourism*, which proposed the ambitious goal of "cultivating tourism as a strategic pillar industry of the national economy and a modern service industry that engenders satisfaction in people." This positioning further clarifies the important status of tourism in China's economic development. In 2011, the State Council designated May 19 as the "China Tourism Day," which earmarked the entry of mass tourism era in China.

The growth stage of China's tourism management undergraduate programs is marked by the following events that are summarized in Table 2.2. The development of China's tourism management undergraduate education at this stage focused on the education and teaching reform. In 2000, the MOE implemented the "New Century Higher Education Teaching Reform Project." In terms of tourism management undergraduate program, Li Tian of Yunnan University, Yong Ma of Hubei University, and Jiang Du of Beijing International Studies University jointly undertook a study *on the teaching reform and development strategy of tourism management education* and co-edited the first book systematically studying Chinese tourism management undergraduate education. The book was published by the Higher Education Press in 2007. In 2001, the "Teaching Reform and Development Strategy of Tourism Management Education in Colleges and Universities in the New Century" Conference was jointly organized by the MOE, CNTA (now the MCT), the Tourism Group of Business Management Teaching Guidance Committee (TGBMTGC), and the Higher Education Press

Table 2.2 Important events in the growth stage of tourism undergraduate education

Year	Events		Implications
2000	*Research on the teaching reform and development strategy of Tourism Management Education* is funded by MOE		Promoting the teaching reform of tourism management undergraduate education
2001	"Teaching Reform and Development Strategy of Tourism Management Education in Colleges and Universities in the New Century" seminar was held at the tourism college of Hainan University		Forming a consensus on teaching reform and textbook construction
2004	The China Tourism Education Summit Forum was held at the Zhuhai Campus of Sun Yat-sen University		Promoting the integration of tourism higher education and tourism industry practice
2008	The China Tourism Education Association was established		Promoting the improvement of talents quality and the construction of tourism faculty
2007	Organized by TGBMTGC	Tourism Management Discipline Construction Work Conference (Shanghai)	Promoting education and teaching reform, project application, textbook construction, and international cooperation
2007		Tourism Management Discipline Construction Seminar between national key colleges and universities (Chongqing)	
2008		China's Tourism Higher Education Review and Prospect Seminar (Shenzhen)	
2009		International Symposium on Globalization of Tourism Education (Shanghai)	
2012	Revising *Undergraduate Programs Catalogue in Colleges and Universities*		Identifying the composition of tourism management-related programs

at the tourism college of Hainan University. This conference discussed two aspects of the teaching reform and textbook construction of the tourism management program in the 21st century (Li & Zhang, 2001). In 2004, the China Tourism Education Summit was held at the Zhuhai Campus of Sun Yat-sen University. Delegates' discussions were around two themes: tourism education and social development; and China's future development direction of tourism higher education from the perspective of the dislocation of supply and demand between tourism education and tourism industry. Both themes are of great significance for promoting the integration of tourism higher education and tourism industry practices. In 2008, the China Tourism Education Association was established to promote the quality improvement of tourism workforce and the construction of tourism faculty. It is also worth noting that since tourism management was classified under Business Management in 1998, the TGBMTGC became the main promoter for the development of the tourism management program through organizing various conferences. With the joint efforts of tourism scholars, colleges and universities, and other institutions, significant breakthroughs were made in the development of tourism management programs. A number of national-level projects on program construction, quality curriculum development, and teaching staff development were funded.

At this stage, the program setup of tourism management undergraduate education had been further adjusted and optimized. Compared to the 1998 version, the *Undergraduate Programs Catalogue in Colleges and Universities* issued by the MOE in 2012 stipulated two changes. First, Tourism Management (1209) as a discipline field was placed under Management (12), in parallel with Business Management (1202). Under the Tourism Management discipline field, three programs were stipulated, namely Hospitality Management (120902), Exhibition Economy and Management (120903), and Tourism Management (120901K).

Since 2013: standardizing stage

The standardizing stage saw a period of constant development of China's tourism industry and was also a period of transformation of tourism management undergraduate education into standardized distinction. Many aspects of the macro environment, such as high-speed railways, internet, blockchain, artificial intelligence, the digital economy, and other scientific technologies have continued to deepen the impact on China's tourism industry. What's more, tourism industry management has become more standardized. The *Tourism Law of the People's Republic of China*, as the first law in the history of China's tourism development, came into effect on October 1, 2013. The law clearly defined the rights and obligations of tourism managers, operators, and tourists. In addition, the status of the tourism industry in the national economy continues to be elevated. The scale of the three tourism markets (domestic, inbound, and outbound) and their contribution to employment have been substantive in the world. Overall, China's tourism industry experienced the transition from mass tourism to quality tourism, which is reflected in the constant standardization of industry management, the continued development of the industry, and the increasing optimization of the industry resources. In this process, tourism management undergraduate education is also moving toward standardization and high quality.

The standardizing stage of China's tourism management undergraduate education mainly includes the following important events as summarized in Table 2.3. In 2013, the MOE issued the *Notice of the MOE on Establishing the Teaching Guidance Committee of Higher Education Institutions from 2013 to 2017*. Since Tourism Management became a discipline field in Management in 2012, the TMTGC was established accordingly. The main responsibility of the TMTGC is to take the consignment of the MOE to carry out research,

Table 2.3 Important events in the standardizing stage of tourism undergraduate education

Year	Events	Implications
2013	TMTGC was established	Carrying out research, consultation, guidance, evaluation, and service of undergraduate teaching in colleges and universities
2018	*National Standards for Teaching Quality of Tourism Management Undergraduate Programs in Colleges and Universities* was announced	Regulating program development and improving teaching quality
2019	"Double Ten-Thousand Plan" for the Construction of First-Class Undergraduate Programs was launched	Promoting Connotative development of tourism management-related programs
2020	*Undergraduate Programs Catalogue in Colleges and Universities (2020 Edition)* was released	Stipulating the attribute and composition of Tourism Management again

consultation, guidance, evaluation, and service to colleges and universities delivering undergraduate tourism programs. The establishment of the TMTGC has provided clear leadership in the standardized development of tourism management-related programs. After its establishment in 2013, the TMTGC compiled the *National Standards for Teaching Quality of Tourism Management Undergraduate Programs in Colleges and Universities* and submitted its final version to the MOE in 2017. In 2018, the TMTGC's standards document was officially announced as one of the components of the *National Standards of Teaching Quality of Undergraduate Programs in Colleges and Universities* and became a guiding document for regulating program development and improving teaching quality. In 2019, the MOE decided to launch the "Double Ten-Thousand Plan" for the construction of First-Class Undergraduate Programs, which aims to build about 10,000 national first-class undergraduate programs and 10,000 provincial first-class undergraduate programs in 2019–2021. Through the organization and selection of the TMTGC, 48 programs were approved to be national first-class undergraduate programs in 2020. The implementation of this development plan will play a leading role in China's tourism management undergraduate education. In future development, the first-class programs and first-class curricula construction plan of tourism management will be further promoted and implemented. At the same time, the *Undergraduate Programs Catalogue in Colleges and Universities (2020 Edition)* further clarifies the structure and components of Tourism Management programs.

Current state

Since the reform and opening-up in 1978, China's tourism management undergraduate education has undergone three stages of inception, growing, and standardizing.

In this process, the program system associated with Tourism Management was formed with increasing numbers of colleges and universities and students.

Program classification

The *Undergraduate Programs Catalogue in Colleges and Universities* issued in 2012 made changes to tourism-related programs. Tourism Management (1209) became a subject under Management (12) and included four programs: tourism management (120901K), hospitality management (120902), exhibition economy and management (120903), and tourism management and service education (120904T). From the perspective of the development stage, China's tourism higher education has experienced the growth process from "tourism economics" program to "tourism management" program and further to the Tourism Management discipline field. It has developed from a single program into a program cluster, indicating that China's tourism higher education is maturing. From the perspective of program composition, the cultivation of a tourism workforce mainly depends on the nature of the tourism management. Hospitality management, exhibition economy, and management rely on related specific industries to prepare management staff with relevant theoretical knowledge and operation skills. Tourism management and service education programs are mainly established in normal (teachers' training) colleges and universities for preparing teaching staff in tourism programs. In terms of learning outcomes, the *National Standards for Teaching Quality of Tourism Management Programs in Colleges and Universities* stipulates the objectives of tourism management-related programs as follows: cultivating application-oriented graduates who master the basic theory, expertise and professional skills of modern tourism management; developing an international perspective; growing management capabilities, service awareness, and an innovative spirit; and engaging in tourism-related operations, management, planning, consulting, training, and education (Tian, Zhao & Guang, 2018).

Development scale

With the development of China's tourism industry and the increasing demand for tourism workers, more and more colleges and universities have established tourism management-related programs. As can be seen in Table 2.4, in 2016, 565 colleges and universities in China established tourism management-related programs. In 2017, this number increased to 569 (Tian & Wu, 2018). In 2018 and 2019, the number of colleges and universities with tourism programs increased to 585 and 597, respectively. In terms of specific programs, although Tourism Management still occupies a dominant position in the discipline field, the pattern of the "one program" structure has gradually been broken, as evidenced in the development trends. The number of colleges and universities with tourism management program shows a slow increase, from 462 in 2016 to 481 in 2019, and the proportion declined from 81.8% in 2016 to 80.6% in 2019. The number of colleges and universities with hospitality management program

Table 2.4 Changes in the number of colleges and universities with tourism
management-related programs

Year	Tourism management-related programs	Tourism management	Hospitality management	Exhibition economy and management	Tourism management and service education
2016	565	462	188	88	17
2017	569	467	204	91	15
2018	585	474	219	102	18
2019	597	481	233	114	18

Source: The catalogue of colleges and universities with tourism management-related programs.

increased significantly, from 188 in 2016 to 233 in 2019, and the proportion
increased from 33.3% in 2016 to 39% in 2019. The number and proportion of
colleges and universities with exhibition economy and management program
also show an upward trend, with the number increasing from 88 in 2016 to 114
in 2019, and the proportion increasing from 15.6% to 19.1%. However, the number of colleges and universities with tourism management and service education
program remained relatively stable throughout this period.

Regional distribution

Currently, colleges and universities with tourism management-related programs
are distributed in 31 provinces of China, showing a stable but uneven distribution
pattern. In 2016, there were 203 (35.93%), 183 (32.39%), and 159 (28.14%) of
these colleges and universities in the eastern, central, and western regions, respectively. In 2019, this distribution changed to 234 (39.2%), 189 (31.66%), and 174
(29.15%). According to these data, it can be found that China's tourism colleges
and universities are still mainly concentrated in the eastern region, and the increase
in the eastern region is greater than that in the central and western regions. When
examined at the provincial level (Table 2.5), there were more than 30 such colleges and universities in Hunan (32), Jiangsu (32), Henan (31), and Hubei (30) in
2016. In 2019, the provinces with more than 30 such colleges and universities are
Henan (35), Hunan (34), Jiangsu (33), Guangdong (32), Sichuan (31), Shandong
(30), and Hubei (30). However, the number of colleges and universities in Hainan,
Tibet, Qinghai, and Ningxia has always been less than or equal to five. Therefore,
it can be concluded that the regional distribution of colleges and universities shows
an uneven pattern from the perspective of regional and provincial levels.

University structure

Colleges and universities with tourism management-related programs in
China have obvious structural characteristics in terms of levels and categories.
As evidenced in Table 2.6, by inspecting the hierarchical structure of colleges and

Table 2.5 Provincial distribution of colleges and universities with tourism management-related programs in 2016 and 2019

2016				2019			
Province	*No.*	*Province*	*No.*	*Province*	*No.*	*Province*	*No.*
Jiangsu	32	Anhui	18	Henan	35	Guizhou	19
Hunan	32	Chongqing	16	Hunan	34	Heilongjiang	18
Henan	31	Shanghai	15	Jiangsu	33	Anhui	18
Hubei	30	Zhejiang	15	Guangdong	32	Chongqing	16
Sichuan	28	Guizhou	15	Sichuan	31	Shanxi	15
Shandong	27	Shanxi	13	Shandong	30	Shanghai	15
Liaoning	24	Beijing	12	Hubei	30	Beijing	14
Hebei	22	Tianjin	11	Hebei	25	Xinjiang	12
Shaanxi	22	Xinjiang	11	Zhejiang	25	Tianjin	10
Yunnan	21	Gansu	10	Liaoning	24	Inner Mongolia	10
Jilin	20	Inner Mongolia	9	Yunnan	24	Gansu	10
Fujian	20	Hainan	5	Fujian	21	Hainan	5
Jiangxi	20	Qinghai	3	Guangxi	21	Qinghai	4
Guangdong	20	Tibet	2	Shaanxi	21	Tibet	3
Guangxi	20	Ningxia	2	Jiangxi	20	Ningxia	3
Heilongjiang	19			Jilin	19		

Source: The catalogue of colleges and universities with tourism management-related programs.

Table 2.6 Structure of universities with tourism management-related programs in 2019

Level	No.	Category	No.	Type	No.
"Double first-class"	55	University	207	Comprehensive	192
Regular	542	College	265	Normal	141
		Independent colleges	125	Engineering	107
				Finance and economics	79
				Agriculture and forestry	34
				Language	20
				Nationality	14
				Medicine	4
				Art	3
				Sport	2
				Media	1

Source: The catalogue of colleges and universities with tourism management-related programs.

universities to explore the overall level, it can be seen that among the 597 universities in 2019, 55 "Double First-Class" universities have tourism management-related programs, accounting for 40.1% of total "Double First-Class" universities and 9.2% of tourism colleges and universities. This structural feature suggests that tourism management undergraduate programs do not receive adequate attention in top universities in China. According to the levels of the institution, enrolment units can be divided into three categories: universities, colleges, and independent colleges. Among the 597 tourism higher learning institutes in 2019, there were 207 universities accounting for 34.7%, 265 colleges accounting for 44.4%, and 125 independent colleges accounting for 21.0%. It is worth noting that colleges and independent colleges are also the main body of new enrolments added in 2018 and 2019. This trend shows that the applied characteristics in the tourism management undergraduate education in China are becoming more and more obvious, which plays an important role in cultivating comprehensive tourism talents. As for the types of institution, the top four types are comprehensive colleges and universities (192), normal (teacher training) colleges and universities (141), polytechnics (107), and finance and economics colleges and universities (79). The types of universities with fewer than five members are specialized in medicine (4), art (3), sports (2), and media (1). Overall, the types of tourism colleges are no longer dominated by traditional comprehensive, normal, and finance and economics colleges and universities (Xiao, 1999). Universities specialized in medicine, sports, arts, and other types of colleges and universities all began to get involved in tourism management programs. The types of colleges and universities with tourism management programs have been diversified.

Student numbers

The number of students choosing tourism management-related programs in China has shown a steady upward trend in recent years (Table 2.7). According to the data released by the MCT, in 2016 China's tourism management-related programs accommodated 58,000 students and the total number in colleges and universities reached 221,000. Among them, tourism management, hospitality management, and exhibition economy and management programs enrolled 34,000, 15,000, and 5,000 students, respectively, and the total numbers of enrolment in these three programs in colleges and universities were 148,000, 46,000, and 18,000. Although the increase of tourism management programs is not as great as that of hospitality management programs (Table 2.4), the enrolment number in tourism management programs is relatively high, which eventually ameliorated the difference in total enrolments between tourism management and hospitality management programs.

Tourism workforce development

After more than 40 years of development, the scale and structure of China's colleges and universities with tourism management-related programs continues to expand and optimize, thereby providing a solid development foundation for the quality development of the tourism industry workforce. To depict the current status of tourism workforce development, this chapter takes 51 "Double First-class" universities with tourism management programs to analyse the forms of enrolment, teaching and learning outcomes, curriculum system, and faculty.

Table 2.7 Student numbers in tourism management-related programs (unit: 1,000)

Year	Tourism management-related program		Tourism management		Hospitality management		Exhibition economy and management		Tourism management and service education	
	Commencing	Total	Commencing	Total	Commencing	Total	Commencing	Total	Commencing	Total
2016	58	221	34	148	15	46	5	18	4	9
2017	59	218	35	150	14	42	5	18	5	8
2018	60	226	36	154	15	44	5	19	4	10
2019	62	232	36	157	15	46	6	20	4	12

Note: The data for 2014–2017 is from the MCT, and the data for 2018–2019 is calculated based on the newly added programs.

Forms of enrolment

In general, the colleges and universities with tourism management-related programs usually recruit students according to specific program or discipline type. For the former, students need to choose the specific program they want when selecting colleges and universities. However, for the latter, students can choose the whole programs cluster when being admitted, and after a year of studying relevant curricula, they can make a choice of which specific program to study further based on their interests and capacities. Therefore, the latter form has become the preferred one for many colleges and universities due to its advantages of considering the comprehensive growth of students.

In the context of higher education in China, tourism management-related programs are still under the disciplines of management, geography, and economics in many universities, resulting in a more diverse form of enrolment. We selected 51 "Double First-class" universities with tourism management-related programs in China, and examined the forms of enrolment according to their enrolment plans and regulations. Judging from the survey results as shown in Table 2.8, "Double First-class" universities mainly recruit students in three disciplines, namely Business Management, Tourism Management, and Sociology. However, there are three additional forms of enrolment in Economic Management, Economic Management Experimental Class, and Liberal Arts Experimental Class, which are trial programs taken by some universities, aiming to improve the quality and efficiency of teaching and learning by integrating relevant programs across different schools and departments within the universities.

As can be seen in Table 2.8, a majority of universities recruit students by disciplines, and this trend is continuing. This is also the mainstream trend in the higher education development in China. Among the "Double First-class" universities with tourism management programs, 28 universities enrol students by specific programs, accounting for 54.9%. Besides, 12 universities recruit tourism students into business management as a discipline, six colleges label tourism management as a discipline for student recruitment, two universities recruit students in economic management, and one university recruits tourism students in sociology. In addition, tourism management as a study subject also appeared in the Economic Management Experimental Class established by two universities and the Liberal Arts Experimental Class established by one university. Overall,

Table 2.8 Enrolment forms of "Double First-class" universities with tourism management-related programs in 2019

Enrolment forms	No.	Enrolment forms	No.
Tourism management (specific program)	28	Economic Management	1
Business Management	12	Economic Management Experimental Class	2
Tourism Management	6	Liberal Arts Experimental Class	1
Sociology	1		

the forms of enrolment into tourism management reflect a diversified and exploratory development trend.

Learning outcomes

The *National Standards for Teaching Quality of Tourism Management Undergraduates in Colleges and Universities* released in 2018 contain detailed regulations on the learning outcomes, curriculum system, faculty, and employment directions of tourism management-related programs. Many universities have also adjusted their programs based on this standard. By analysing the learning outcomes of 51 "Double First-class" universities, it was found that the aim of existing programs has designated the learning outcomes in three aspects: ability requirements, employment direction, and talent positioning.

The learning outcomes of the tourism management program mainly include theoretical knowledge, basic skills, innovative ability, international perspective, and so on, and there are no obvious differences between universities. Employment outlets usually involve the four types of tourism organizations, namely tourism enterprises and institutions, tourism education, scientific research institutions, and secondary or tertiary vocational tourism schools. Specifically, about 58% of colleges and universities focus on tourism administrative departments and enterprises, and 42% of colleges and universities attach importance to tourism education and research institutions and secondary or tertiary vocational tourism schools. For the positioning of talents, there are 16 universities that propose to cultivate applied talents and compound talents and 18 universities aiming to develop middle and senior managers. It is worth noting that Dalian Maritime University and Beijing Sport University have incorporated their own disciplinary advantages into their learning objectives, relying on the cruise yacht industry and sports tourism to train tourism managers in these areas, and thus form their unique program characteristics.

Although the number of "Double First-class" universities with hospitality management, exhibition economy and management programs is relatively small, there are obvious different learning objectives/outcomes formulated in these two programs, which are mainly reflected in the directions of employment. The employment direction of hospitality management is mainly concentrated in the hotel accommodation sector. The employment direction of the exhibition economy and management program is more diversified, including the fields of conferences, exhibitions, festivals, etiquette, events, performances, cultural communication, and tourism.

Curriculum system

In China's higher education context, the curriculum system generally includes four parts: liberal art education courses, basic disciplinary courses, core program courses, and practice-based courses. Among them, core program courses, as a concentrated manifestation of program-related theoretical knowledge, are the most important part of the higher education curriculum system. The *National Standards for Teaching Quality of Tourism Management Undergraduates in Colleges and Universities* stipulates a "4 + 3" professional curriculum system for tourism management-related programs, including four core courses and three program courses. Among them, the four core courses are Introduction of Tourism, Hospitality Industry, Tourism Destination Management, and Tourism Consumer Behaviour. For tourism management, the program courses are Tourism Economics, Tourism Resource Management, and Tourism Law. The program courses of hospitality management are Introduction to Hospitality Management, Hotel Operation Management, and Hotel Customer Management. Finally, the program courses of exhibition economy and management are Introduction to Exhibition Management, Exhibition Project Management, Exhibition Planning and Design.

As illustrated in Figure 2.1, of the 45 "Double First-class" universities with available course data, 39 universities offer courses related to Introduction of Tourism, four offer courses related to Hospitality Industry, eight offer courses related to Tourism Destination Management, 12 offer courses related to Tourism Consumer Behaviour, 33 offer courses related to Tourism Economics whilst 28 offer courses related to Tourism Resource Management, and 21 universities offer courses related to Tourism Law. However, there are only three universities offering all seven courses at the same time, indicating that the current curriculum system of China's tourism management programs needs to be further improved.

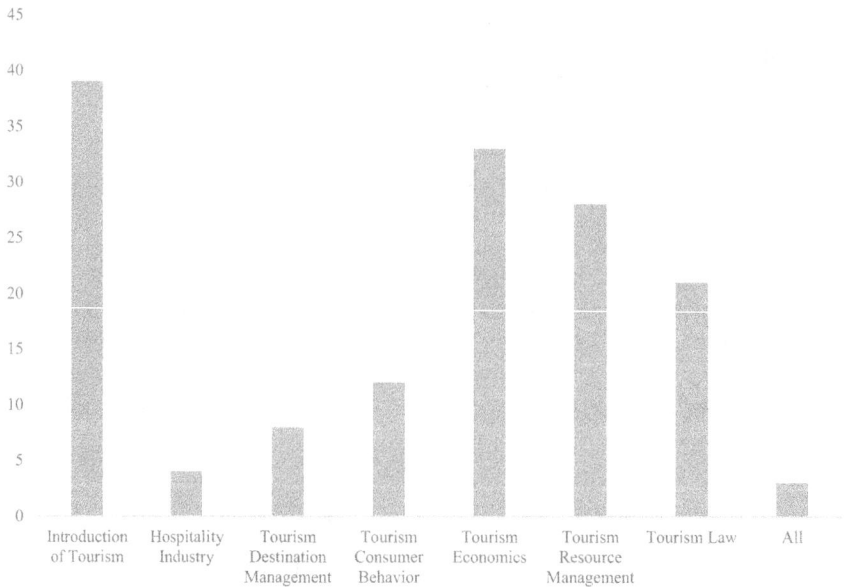

Figure 2.1 The professional curricula of "Double First-Class" universities with tourism management programs.

Through the analysis of data from six "Double First-class" universities with exhibition economy and management courses, it was found that only one university offer four core courses, and five universities offer three program courses. This pattern shows that the popularity of Tourism Management core courses is not commonly found in the exhibition economy and management program.

Faculty

The *National Standards for Teaching Quality of Tourism Management Undergraduates in Colleges and Universities* describes the scale and structure of teachers in tourism management-related programs, requiring that the number of teachers of each program should be no less than ten, the proportion of lecturer and above positions should be no less than 80%, and the proportion of senior titles should be no less than 20%. By visiting the official website of the "Double First-class" universities with tourism management-related programs to survey the number of professors, associate professors, and lecturers, 39 universities were finally selected in the analysis. As shown in Figure 2.2, only seven universities have fewer than ten teachers; most universities basically meet the requirements of national standards. In addition, the number of teachers in 24 universities is between 10 and 19 and the number of teachers in three universities (Ningbo University, South China University of Technology, Yunnan University) is between 20 and 29. Four universities (Hainan University, Nankai University, Sun Yat-sen University, and Sichuan University) have a faculty of more than 30 academic staff, with Hainan University ranking the first, with 53 full-time academic staff.

From the perspective of faculty structure, there are 24 universities with more than 20% of teachers with senior titles, and the top 10 universities listed in

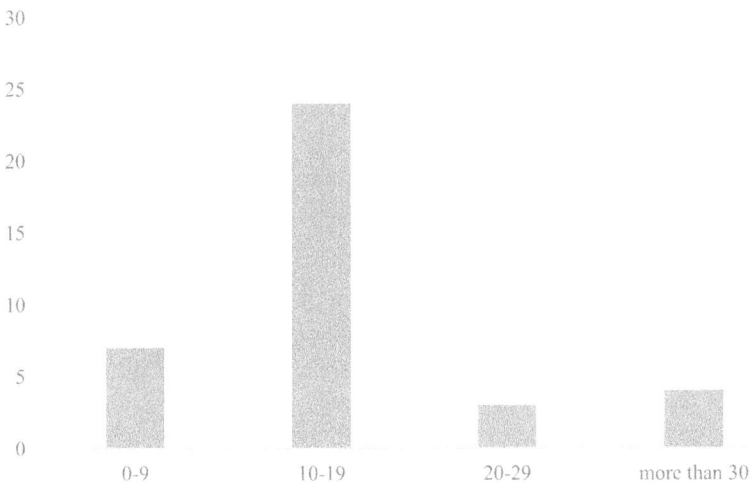

Figure 2.2 The faculty number of "Double First-Class" universities with tourism management programs.

Table 2.9 Faculty structure of the top ten universities in tourism management-related programs

Name	No.	Professor (%)	Associate Professor (%)	Lecturer (%)
Hainan University	53	28.3	32.1	39.6
Nankai University	37	21.6	48.6	29.7
Sun Yat-sen University	35	34.3	57.1	8.6
Sichuan University	30	40.0	33.3	26.7
Yunnan University	27	25.9	33.3	40.7
Ningbo University	23	30.4	39.1	30.4
South China University of Technology	21	23.8	38.1	38.1
Guangxi University	18	50.0	33.3	16.7
Hunan Normal University	17	29.4	41.2	29.4
Chengdu University of Technology	17	23.5	29.4	47.1

Source: Official website of each university.

Table 2.9 are: Guangxi University (50%), Beijing Jiaotong University (42.9%), Xiamen University (42.9%), Sichuan University (40), Jinan University (40%), Shandong University (38.5%), Liaoning University (36.4%), Sun Yat-sen University (34.3%), Southwest University of Finance and Economics (33.3%), and Nanjing Forestry University (33.3%). There are 36 universities with more than 40% of teachers with associate professor title. The top ten universities are: Beijing Jiaotong University (100%), Shandong University (100%), Xiamen University (100%), China University of Geosciences (Wuhan) (100%), Sun Yat-sen University (91.4%), Beijing Forestry University (90.9%), South China Normal University (90.9%), Ocean University of China (84.6%), Southwest University of Finance and Economics (83.3%), and Guangxi University (83.3%). In general, the faculty structure of the "Double First-Class" universities basically meets the requirements of the *National Standards for Teaching Quality of Tourism Management Undergraduates in Colleges and Universities.* However, it should be noted that because these universities are mostly research-intensive universities, the faculty profile is stronger compared to other universities.

Quality management

Tourism management undergraduate education has been in a volatile development process for a long time since its birth. Many problems existed in the areas of discipline construction, program development, talent training, and so on, which significantly slowed down the development for the programs' quality management. What is gratifying is that some colleges and universities rely on their disciplinary advantages to build their tourism management programs, accumulating significant amounts of development experience, thus qualifying for international certification, such as the UNWTO Tourism Education Quality Certification (UNWTO-TedQual). However, this international certification is only applicable to a small number of colleges and universities and has very strict requirements. For most other colleges and universities in China, quality management based on the *National Standards for Teaching Quality of Tourism Management Undergraduates in Colleges and Universities* can be considered in the early stages of development.

UNWTO-TedQual is a global TedQual project supported by the UNWTO Themis Foundation, which is responsible for the implementation of research projects, education, and training of the UNWTO as well as evaluation and certification of the quality of tourism education institutions worldwide. At present, the colleges and universities that have obtained this certification in the tourism management-related programs in mainland China include Sun Yat-sen University (in Tourism Management, Exhibition Economy and Management), Beijing International Studies University (in Tourism Management, Hospitality Management, Exhibition Economy and Management), Xi'an Eurasia University (Tourism management), South China Normal University (Tourism Management, Hospitality Management, Exhibition Economy and Management), Jinan University (Tourism Management, Hospitality Management, Exhibition Economy and Management), Huangshan University (Tourism Management, Hospitality Management,) and Zhuhai College of Jilin University (Tourism management).

The formulation of *National Standards for Teaching Quality of Tourism Management Undergraduates in Colleges and Universities* ("national standard") aims to clarify the basic requirements of undergraduate programs in Tourism Management, guide the direction of program development, encourage colleges and universities to identify their advantages and characteristics, improve the quality of program teaching, and encourage colleges and universities to serve economic and social development. In terms of the content, the "national standard" mainly includes six parts: objectives, training specifications, curriculum system, faculty, teaching conditions, and teaching effects. These are also the main components in higher education to build a quality assurance system. Since its release in January 2018, many colleges and universities in China have adjusted their teaching programs and built a quality management system based on this standard.

Issues and challenges

After more than 40 years of development, China's tourism management undergraduate education has reached significant milestones and made a lot of achievements. Meanwhile, the development of tourism management undergraduate education still faces a series of problems and challenges. The most salient issues are discussed below.

Intensifying competition

Although the discipline status of tourism management has been improved since 2012, tourism management-related programs are still under the auspices of business management in many colleges and universities as a secondary discipline (Yin

& Meng, 2018). In this context, even if some colleges and universities have already started to recruit students by discipline type, it is still a common situation for colleges and universities to recruit students in tourism management-related programs through the business management route. For colleges and universities that recruit students according to discipline type of Tourism Management, there won't be a significant degree of competition in appealing to students between specific programs for the sake of similar knowledge system. However, for colleges and universities that recruit students into Business Management or other discipline types, tourism management programs will face fierce competition. Compared with other programs, tourism management-related programs have obvious problems such as unclear knowledge boundaries and low skill requirements. Meanwhile, the tourism industry also has practical problems such as high turnover rates and low average wages (Gu, Kavanough, & Yu, 2007). These detrimental situations will adversely affect student choice of programs, which will weaken tourism management-related programs.

Weak features

After analysing some of the "Double First-class" tourism programs in universities, it is found that many universities show obvious homogeneity in learning goals, curriculum systems, faculty, and so on, which makes tourism programs lack in vitality and unique features. In terms of the learning objectives, most universities show no distinctions in ability requirements, employment direction, and talent positioning. Most programs do not have distinctive subject advantages and have not incorporated industry development needs in their curricula. It is worth noting that some universities have outstanding performance in this respect, such as Dalian Maritime University and Beijing Sport University. However, in terms of the curriculum system, most universities ignore the development of specialty courses while designing the teaching content around the core courses. In terms of faculty, most senior faculty members in the field of tourism management are from geography, economics, and other disciplines. This situation can essentially confine some teachers' understanding of the tourism management-related programs to be viewed through the prisms of their original subject backgrounds, which may influence the construction and transfer of theoretical knowledge. Of course, it should be noted that the training of China's Tourism Management professionals is still in the stage of dynamic adjustment and improvement, which is part of a long-term process.

Program development plans

The connotative development of China's higher education essentially consists of three interconnected plans, namely the "Double First-Class Plan," the "Double Ten-Thousand Plan," and the "Six Excellence and One Brilliance Plan 2.0." Among them, the "Double Ten-Thousand Plan" and the "Six Excellence and

One Brilliance Plan 2.0" are directly related to tourism management-related programs. Further development of tourism management undergraduate education needs to be closely related to these plans.

Building first-class programs

In April 2019, the MOE issued the *Notice on the Implementation of the "Double Ten-Thousand Plan" for the Construction of First-Class Undergraduate Programs*, which aims to build about 10,000 national first-class undergraduate programs and 10,000 provincial first-class undergraduate programs in 2019–2021. In December 2019, the MOE announced 4,054 national first-class trial programs and 6,210 provincial first-class trial programs. Among these, 48 are national first-class Tourism Management programs, including 41 in tourism management, 5 in hospitality management, and 2 in exhibition economy and management.

Teaching and learning approach

The teaching and learning (T&L) approach is a theoretical model and operation style for the whole process of learning formulated for the realization of specific learning objectives (Dong, 2012). It is a programmatic approach for program construction, and includes three core elements. First of all, the most important issue is to define the learning outcomes or graduate qualities. This can be based on the learning objectives stipulated in the *National Standards for Teaching Quality of Tourism Management Undergraduates in Colleges and Universities* and the advantages of the university. On this basis, practical, innovative, and unique learning objectives can be formulated. Second, teaching programs should be designed to clarify the learning specifications, curriculum composition, teaching staff, teaching management, evaluation system, and other modules. Finally, teaching plans should be developed to implement the programs in specific dimensions such as time, space, tasks, and assessment. By formulating learning goals, teaching programs, and teaching plans, the advantages and characteristics of the T&L approach can be established.

Faculty

The faculty is at the core of program construction and undertakes the dual roles of spreading knowledge and organizing teaching. In terms of disseminating knowledge, the size and qualification structure of the faculty directly affect the quality of teaching programs. Teaching staff also take the primary role in teaching delivery. The *National Standards for Teaching Quality of Tourism Management Undergraduates in Colleges and Universities* set requirements on the scale, structure, background, level, teaching, and development planning of a university faculty; these requirements provide useful guidelines for faculty development.

Practicums

Practicums are an important channel for applying theoretical knowledge to industry practice. Therefore, it is necessary to build high-quality practicums on and off campus. To build an on campus practicum, it is necessary to improve the applicability of courses such as tourism planning, tourism information technology, and tour guide services through campus seminar rooms, museums, and outdoor courses, with the premise of comprehensively considering the content of Tourism Management courses. In terms of off campus practicums, multiple internship programs in attraction management, folklore culture preservation, smart technology, and investment operation can be developed with industry collaborations.

International collaboration

International collaboration is an important way to improve the quality of teaching programs. There are three strategies to promote the construction of first-class undergraduate programs. The first is to expand channels for international cooperation (Lin, 2019). Collaborative models such as mutual recognition of credits, cosponsoring, cooperative cultivating, and overseas internships can be adopted. The second is to highlight the various characteristics. Different regions and types of tourism colleges and universities can adopt different ways of cooperation. Developed regions can explore the model of cosponsoring and overseas internships, while developing regions are mainly based on the students exchange and cooperative cultivating. The third is to promote the international exchange of teachers. Teachers can be exchanged to foreign universities for further studies as visiting scholars.

Designing first-class curricula

In October 2019, the MOE issued the *Implementation Opinions on the Construction of First-Class Undergraduate Courses*, which aims to build about 10,000 national first-class undergraduate courses and 10,000 provincial first-class undergraduate courses. Currently, first-class courses are still in the selection process. Promoting the construction of first-class courses of tourism management-related programs can be carried out in terms of curriculum content, teaching forms, teaching methods, and assessment methods.

Curriculum content

Course content should be designed to achieve the goal of student knowledge acquisition and skill development. In operation, the course content design must rely on the "4 + 3" program system, reflecting the development of the tourism industry and using specific practical cases. At the same time, the course content and cases should focus on the integrated development of culture and tourism.

Teaching forms

The teaching forms are mainly manifested in the various first-class courses, including online courses, offline courses, online and offline mixed courses, virtual simulation experiment teaching courses, and social practice courses. It should be noted that the prescribed curriculum system of "4 + 3" is also the basis for the creation of tourism management-related program courses.

Teaching methods

The purpose of designing teaching methods is to enable students to learn how to use theoretical knowledge to explain tourism development practices and cultivate industry perception. Modern scientific and technological methods should be incorporated in the design of teaching methods to guide students to actively attend to various tourism phenomena generated by the tourism industry through group discussions, debating, and field research.

Assessment methods

Procedural assessment methods instead of result-based assessment should be implemented to focus on student enthusiasm and initiative in the learning process. In terms of assessment basis, teachers should actively carry out group discussions, outdoor activities, field research, and make appropriate records to systematically evaluate student learning effects.

Cultivating first-class talents

In April 2019, the "Six Excellence and One Brilliance Plan 2.0" was launched in Tianjin to promote the construction of new programs in engineering, medical sciences, agricultural sciences, and liberal arts, and further improve the ability of colleges and universities to serve economic and social development. The construction of new liberal arts in tourism management-related programs needs to be carried out from the integration of technology and teaching, interdisciplinary platform, and cooperative education mechanism.

Integration of technology and teaching

The technology-teaching integration model is a measure that integrates modern science and technology into the teaching process to improve the teaching level and talent quality. In terms of teaching form, by applying big data, artificial intelligence, virtual reality, and other technologies to build smart classrooms and smart laboratories, students can learn the wide application of technology in modern tourism. Besides, it is also possible to promote the transformation of teaching activities through field investigations. Taking the application of regional digital tourism as an example, students can be guided to pay attention to the integration trend of technology and tourism.

Interdisciplinary platform

The construction of an interdisciplinary platform is an initiative to develop interdisciplinary talents based on the needs of theoretical knowledge and curriculum system of the tourism discipline. To build an interdisciplinary platform for Tourism Management, it is necessary to promote the integration of tourism management-related programs with business management, industrial economics, human geography, consumer psychology, computer science, information technology, and other disciplines to cultivate comprehensive talents. The focus of platform construction mainly includes the interdisciplinary curriculum system, teaching team, research topics, and so on.

Cooperative education mechanism

The cooperative education mechanism is an operation mode in which colleges and universities, governments, enterprises, research institutions, and so on are the main bodies to jointly cultivate talents. From the perspective of the composition system, the cooperative education mechanism includes four aspects: goals, teachers, resources, and management. The purpose of establishing the target cooperation mechanism is to revise the teaching plan to clarify employment skills and quality requirements for the workforce. It is, therefore, important to establish a faculty collaboration mechanism to promote communication and cooperation between full-time teachers in the school and part-time teachers outside the school. It is also recommendable to develop a resource sharing mechanism to promote the sharing of education and practice resources among students, teachers, and employers.

Conclusions

Since 2019, a series of programs in the field of higher education in China have focused on undergraduate education, program construction, and teaching and learning approaches. Under this premise, this chapter takes the tourism management undergraduate education as the focus and analyses it in the aspects of historical review, current state, talent training, issues and challenges, and construction plan.

Three stages: inception, growing, and standardizing

Based on the adjustment of the tourism management programs, the period between 1979 and 1998 was defined as the inception stage and included the important events of the establishment and the introduction of the tourism management undergraduate programs. The period of 1999–2012 is characterized as the growth stage. The most notable and meaningful event in this period was the upgrade of tourism management from a single program to a program cluster, with a program system basically formed. Tourism management undergraduate education came into the standardizing stage in 2013. Symbolic events include the establishment of the TMTGC and the release of the *National Standards for Teaching Quality of Tourism Management Undergraduates in Colleges and Universities.*

Current state

In terms of program classification, tourism management undergraduate education includes four programs: tourism management, hospitality management, exhibition economy and management, and tourism management and service education. In 2019, there were 481 colleges and universities with tourism management, 233 with hospitality management, 114 with exhibition economy and management, and 18 with tourism management and service education. In terms of development scale, there were 597 colleges and universities offering tourism management-related programs in 2019. The number of commencing students enrolled on these programs is 62,000 and the total number of students at colleges and universities is 232,000. In terms of regional distribution, the respective distributions of colleges and universities in the eastern, central, and western regions in 2019 were 234 (39.2%), 189 (31.6%), and 174 (29.15%).

Talent training

The forms of enrolment into tourism management-related programs include enrolment by specific programs and enrolment by discipline fields. In terms of learning goals, the three aspects of ability requirements, employment direction, and talent positioning should be considered. The curriculum system includes four core courses (Introduction of Tourism, Hospitality Industry, Tourism Destination Management, Tourism Consumer Behaviour) and three specialty courses. The total number of teachers should be no less than ten, the proportion of lecturers and above should be no less than 80%, and the proportion of senior titles should be no less than 20%. For quality management, some colleges and universities have been certified by the UNWTO-TedQual, while most domestic colleges and universities established a quality management system based on the *National Standards for Teaching Quality of Tourism Management Undergraduates in Colleges and Universities.*

Issues and challenges

Enrolment based on the discipline fields increases the degree of competition among kinship programs. The homogeneity of colleges and universities in the field of talent training also weakens program construction. The in-depth integration of culture and tourism and the limitations of program settings in colleges and universities have brought challenges to the collaborative training of cultural and tourism talents.

References

Dong, Z.F. (2012). On the concept and constitution of university's cultivating pattern. *University Education Science*, 3, 30–36.

Gu, H.M., Kavanaugh, R.R. & Yu, C. (2007). Empirical studies of tourism education in China. *Journal of Teaching in Travel & Tourism*, 7(1), 3–24.

Li, D.M. & Zhang, L.L. (2001). Integrate tourism education and strengthen the talent project-summary of Seminar on teaching reform and development strategy of tourism management program in colleges and universities. *Tourism Tribute*, 16(3), 39–40.

Lin, J. (2019). First-class undergraduate education: Construction principles, construction emphases and the guarantee mechanism. *Tsinghua Journal of Education*, 40(2), 1–10.

Tian, L. & Wu, X.Z. (2018). Annual report of tourism management undergraduate education. In China Tourism Education Association (Ed.) *China tourism education blue book* (pp. 29–49). Beijing: China Tourism Press.

Tian, L., Zhao, S.H. & Guang, Y.J. (2018). The national standards for teaching quality of tourism management undergraduates in colleges and universities. In Higher Education Teaching Guidance Committee of the Ministry of Education (Ed.) *The national standards for teaching quality of undergraduate programs in colleges and universities* (pp. 896–900). Beijing: Higher Education Press.

World Tourism Organization (UNWTO). (2019). *International Tourism Highlights, 2019 Edition*. Retrieved August 28, 2019 from https://www.unwto.org/publication/international-tourism-highlights-2019-edition.

Xia, J.C. & Xu, J.H. (2018). Reform and opening-up of tourism in China from 1978 to 2017: Retrospects and prospects. *Research on Economics and Management*, 39(6), 3–14.

Xiao, H.G. (1999). Tourism education in China: Past and present. *Asia Pacific Journal of Tourism Research*, 4(2), 68–72.

Zhang, W. & Fan, X. (2005). Tourism higher education in China. *Journal of Teaching in Travel & Tourism*, 5(1–2), 117–135.

Yin, Z. & Meng, F. (2018). Tourism higher education in China: Profile and issues. In J. Zhao (Ed.) *The hospitality and tourism industry in China: New growth, trends, and developments* (pp. 241–261). Palm Bay, FL: Apple Academic Press.

3 China's higher education in hospitality management

Bin Li and Huimin Gu

Introduction

China's hospitality industry has grown rapidly in recent years. By the end of 2019, the number of star-rated hotels in China had reached 10,130, and the scale of hotel groups was expanding swiftly.[1] The 2018 China Hotel Management Companies (Groups) Development Report indicated that the 63 largest hotel groups in China were operating 28,950 hotels containing 3,111,621 total rooms.[2] Along with the booming number of hotel groups, China's hospitality industry has also expanded overseas; the number of hotel groups "going out" has increased substantially. For example, in the *Hotels* magazine's 2019 ranking, six Chinese hotel groups—Jinjiang International, Huazhu Group (formerly China Lodging Group), BTG Hotels Group, Green Tree Hospitality Group, Dossen International Group, and Qingdao Sunmei Group—ranked among the top 20 hotel groups globally. China has simultaneously boosted its efforts to leverage the layout of third- and fourth-tier cities to promote hospitality groups' domestic expansion. The rise of a new business format, hotels based on innovative technology (e.g., mobile internet technology, cloud computing, and big data), has also had a noteworthy impact on the traditional accommodation industry.

Hospitality management became a subfield of tourism-related higher education in 1980. China's Ministry of Education later approved this domain as a university teaching program in 2003. The hospitality management program began recruiting students at 25 universities in 2004 and has since mushroomed: in 2017, 220[3] colleges in China offered a hospitality management major—a 24% jump over 2016. This figure has remained stable since. As of this writing, 14% of universities in China offer a hospitality management major. The vast majority (92%) of institutions with hospitality management programs is second- and third-tier universities; only 7% of such programs are hosted at top Chinese universities.

In 2018, the Ministry of Education of China promulgated the *Six Outstanding and One Top* plan. Under this plan, the "new liberal arts subject" was proposed as a subject area combining advanced science and technology with traditional liberal arts (Chen & He, 2020; Li, 2020). The "new business subject" was also developed. Amid the era of the digital economy, traditional industrial

DOI: 10.4324/9781003004363-3

organizations, business models, and management models must be updated. In particular, hotel management subjects and talent-training models need to be re-vamped. The hospitality industry requires more international, well rounded, and applied talent as well. These demands call for industry practitioners who possess specialized education in hospitality management.

Literature review

In the past 30 years, China's higher education programming has aligned with international standards and made great progress. Yet the country's higher education sector continues to encounter new challenges, mainly reflected in the effects of a changing population structure, swift economic growth, the massification of higher education, and mismatches between university teaching and workplace requirements (Bie & Yi, 2014). These circumstances point to a need for China's higher education sector to adopt new development models, adapt to social evolution, meet societal needs, and optimize curricula to ensure balanced educational programming. Many scholars have presented proposals tailored to China's educational development needs. Suggestions have entailed the modernization of higher education, higher education management, and the growth of "new liberal arts subjects" (Chen & Zhan, 2009; Dong et al., 2013; Liu, 2014).

As a new and distinctive subject area in China's higher education landscape, tourism and hospitality education has become a hot topic in academia. China's tourism industry and related education emerged relatively late. Researchers have often taken lessons from Western countries such as Australia, Europe, and the US (Gu et al., 2003; Smith & Ding, 2001; Wu & Li, 2005; Xu & Zhang, 2004). But given distinct cultural features and degrees of economic development, it remains challenging to incorporate Western countries' hospitality education experiences into China's educational system. Studies have highlighted various concerns in tourism education studies (Goodman & Sprague, 1991; Hemmington, 1995), such as those involving curriculum design, principles, and methods applied to China's higher education sector in tourism and hospitality (Qin, 2004; Tribe, 2001; Zou et al., 2002).

Despite rapid growth in the number of China's hospitality programs, quality development poses an obstacle compared with other traditional business disciplines such as accounting, finance, and marketing. More attention should be given to hospitality higher education, with greater investment in educational resources and the development of teaching faculty (Yu, 2008). Relatively little research has considered hospitality management higher education; however, many case studies have detailed world-famous education models exemplified by Purdue University, University of Houston's Conrad N. Hilton College of Hotel and Restaurant Management, and Lausanne Hospitality Management College. The development of professional education at these colleges serves as a reference for the development of hospitality management in China (Chen & Dellea, 2015; Gu et al., 2003; Liu & Enz, 2006; Penfold et al., 2012). Regarding professional education models and talent training in hospitality management in China, close consideration should be given to the nature of practice-based teaching. Pertinent areas of interest include innovative teaching models, fostering international perspectives among students, and developing a balanced knowledge structure (Bao & Zhu, 2008; Lam & Xiao, 2000; Peng, 1999; Zou et al., 2002). As an exception in the literature, Dai (2005) systematically studied the talent cultivation, research methods, and teaching approaches of postgraduate education in hospitality management.

Several open questions remain in this realm, especially regarding employee turnover and graduates' willingness to work in this industry from an educational perspective. This chapter, based on an empirical study, provides a comprehensive review of hospitality management programs in China in terms of higher education institutions, faculty, and curricula.

Research methods

We used a sample questionnaire survey method to gather data. The questionnaire design was based on main indicators in the "Undergraduate Teaching Quality Observation Point" of the Higher Learning Institutes in China and the "List of Economics and Management Subjects" issued by the Ministry of Education's Degree and Postgraduate Development Center. The 40 survey questions covered several topics related to the scale of universities offering hospitality management programs: faculty; teaching, learning, and research; social services; international cooperation; and demographic characteristics.

The questionnaire was distributed on-site and online. On-site distribution was mainly intended for teachers attending the National Hospitality Management Training Program hosted by China Tourism Education Association and Beijing International Studies University in July 2016. One hundred questionnaires were distributed, 50 of which were returned. The online survey was conducted through WeChat groups comprising tertiary-level hospitality management teachers in China. The groups included 382 active hospitality teachers from 81 universities nationwide. To avoid repetitive answers (as some intended respondents may have received the questionnaire on-site and online), we only retained on-site responses if the same respondent completed both the on-site and online questionnaire. We received 81 valid responses from 64 universities, representing 29% of the higher education institutions offering hospitality management programs in China.

Data analysis and results

Institutional features

Geographical distribution

As shown in Figure 3.1, the number of universities offering hospitality management programs in first-tier cities represented only 4% of schools listed in the survey. Most universities (80%) providing such programs were in second- and third-tier cities, with 16% in fourth-tier cities. The distribution of higher education

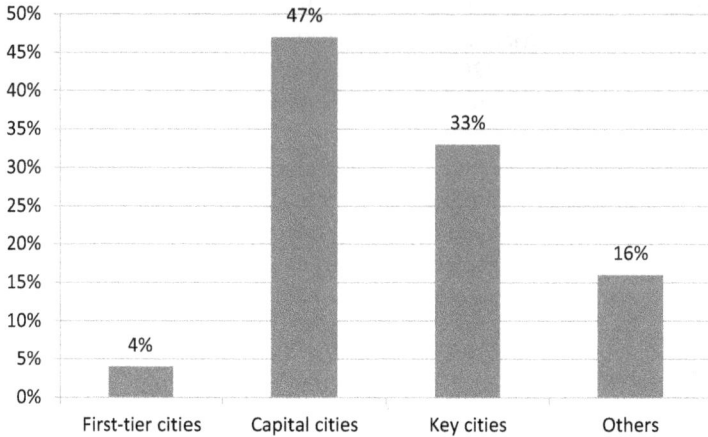

Figure 3.1 Geographic distribution of hospitality management programs.

institutions offering hospitality management programs did not appear to match geographic needs for industry practitioners. More specifically, first-tier cities generate a large proportion of hospitality jobs, yet the majority of universities with hospitality programs were in second- or third-tier cities. Notably, this situation is changing as hotel chains expand to third- and fourth-tier cities in China.

Class distribution

In China, universities were initially classified into first, second, and third tiers when recruiting students based on their national university entrance examination scores. First-tier universities are thought to be of comparatively better quality. Our survey sample contained 17 first-tier universities, 43 second-tier universities, and four third-tier universities.

Organizational structure

Regarding program structure, as shown in Figure 3.2, 41% of hospitality management programs were affiliated with a broad tourism management subject; only 8% were housed in independent hospitality management schools. These independent schools were the first to introduce such hospitality management programs and are thus more comprehensive and competitive. Nearly 63% of colleges/universities currently have independent hospitality management programs. With further industry development and ongoing discipline-based improvements, more of these programs are expected to emerge.

Education levels

Based on our sample, first-tier universities enrolled the lowest number of students while second- and third-tier universities accounted for most undergraduate

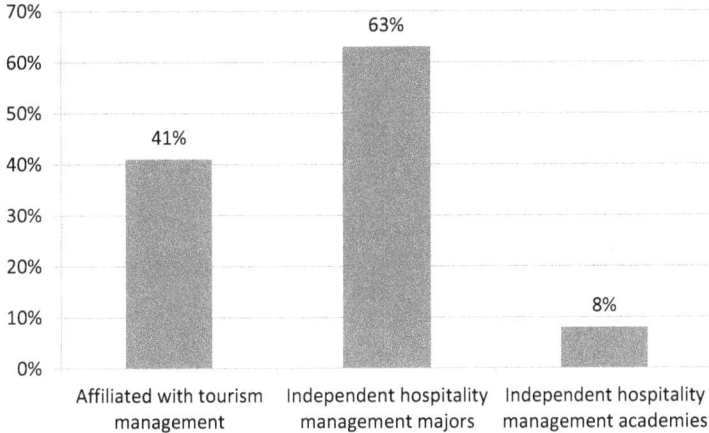

Figure 3.2 Major affiliations of hospitality management colleges.
Percentages overlapped across majors in some cases.

Table 3.1 Admission numbers to hospitality management schools

	Admissions		First-tier colleges		Second-tier colleges		Third-tier colleges	
	Total	Average	Total	Average	Total	Average	Total	Average
Bachelor's program	7,067	110	1,150	72	4,637	110	1,280	427
Master's program	128	9	88	11	110	8	0	0
PhD program	9	2.3	9	2	0	0	0	0

enrollment. Third-tier universities constituted 6% of all enrollments but enrolled as many as 427 students per university on average (Table 3.1). In terms of education level, most hospitality management students in China were pursuing a bachelor's degree; only 128 were in a master's program. Only four universities offered a doctoral degree in hospitality management at the time of data collection. Compared with programs in tourism management, those in hospitality management were in early development with a relatively weak foundation for graduate education. Even some esteemed universities bearing high academic standards and a strong social reputation did not meet qualifications to train doctoral students.[4]

Graduate program characteristics

As illustrated in Figure 3.3, 75% of universities positioned their programs as application-oriented, while 24% programs were application- and research-oriented. Only Sun Yat-sen University's hospitality management program was exclusively research-oriented.

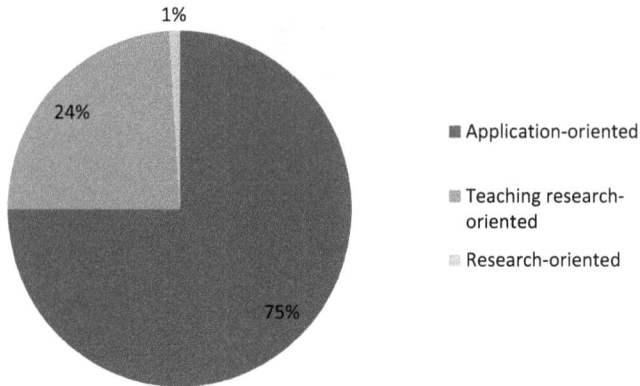

Figure 3.3 Types of hospitality management majors.

Teaching staff

The overall student-to-teacher ratio in these 64 universities was 41:1, much higher than the national requirement of 18:1. University teachers in hospitality management programs generally had a higher teaching load than teachers in other programs and did not participate in industry practice, research, or teaching reform activities. As shown in Table 3.2, first-tier university programs had 11 faculty members on average. Second- and third-tier programs had 12 and 9 members, respectively. Among all teaching staff, about 10% were full professors, 34% were associate professors, and 52% were lecturers. Of the 697 survey respondents, 187 held a doctoral degree and 135 had overseas study experience. Faculty members' supervisory duties differed by university tier: within hospitality management master's and doctoral programs, 69% of the teaching staff at first-tier universities supervised master's students, while 31% of the teaching staff in second-tier universities did so. Slightly more than half (53%) of hospitality management faculty at first-tier universities and 47% at second-tier universities supervised PhD students. Table 3.2 provides more details.

In China, awards are representative of teaching quality and function as an essential quality assessment indicator of university work in the country. Among all survey respondents, 2% ($n = 12$) had won a national award and 5% ($n = 35$) had earned provincial or ministerial awards. A lack of high-quality teachers and talent in hospitality management programs further complicates these programs' development in China.

Among the teachers surveyed, 69% held a master's degree as their highest education level, 15.5% held a doctorate, and 15.5% had a bachelor's degree. Most possessed a background in management or other disciplines. For example, 34% and 12% of faculty had specialized in human geography and physical geography, respectively; only 5% had a dedicated background in hotel management.

Table 3.2 Faculty structure

Faculty	Total	First-tier colleges			Second-tier colleges			Third-tier colleges		
		Total	Proportion (%)	Average	Total	Proportion (%)	Average	Total	Proportion (%)	Average
Faculty	697	169	24	11	501	72	12	27	4	9
Professor	67	27	40	2	39	58	1	1	2	0.3
Associate professor	240	61	25	4	173	72	5	6	3	2
Lecturer	361	82	23	6	259	72	6	20	5	7
Women	487	98	20	7	374	77	9	15	3	8
PhD	187	82	43	6	102	55	3	3	2	2
Overseas education	135	59	44	4	72	53	2	4	3	2
Master's student supervisor	129	89	69	6	40	31	2	0	0	0
PhD student supervisor	15	8	53	0.6	7	47	0.3	0	0	0
National Talent Award title	12	2	17	0.1	10	83	0.2	0	0	0
Provincial Talent Award title	35	13	37	0.8	22	63	0.5	0	0	0

Table 3.3 Teachers with more than one year of industry experience

	Total	First-tier colleges	Second-tier colleges	Third-tier colleges
Total number of teachers	697	169	501	27
One year + working experience	319	44	261	14
Average years of industry experience	5	3	6	5
Proportion	46%	26%	52%	50%

The teaching staff in hotel management programs were typically either young or middle aged: 38% were below 35, 41% were between 36 and 45, and 21% were over 45.

In terms of industry work experience, about 46% of the teaching staff possessed more than one year of industry experience. First-tier university teachers generally possessed less industry experience than their second- and third-tier university counterparts (see Table 3.3).

Teaching quality assessment

The Undergraduate Teaching Assessment Guidelines from China's Ministry of Education outline several key performance indicators denoting quality programs: the number of major-related courses, bilingual courses, elite courses, labs, experimental courses, high-quality textbooks published, teaching awards, student awards, internships, further study upon graduation, employment, and starting salary. The higher the percentage of these attributes, the better the program quality. Table 3.4 illustrates the current quality situation in China.

Among schools participating in the survey, each had ten courses related to hotel management on average. Some institutions, such as Beijing International Studies University, Wuyi University, and Northwest Normal University, offered more than 22 major-related courses.

The average proportion of bilingual courses was approximately 15% but could exceed 50% in foreign language-featured universities (e.g., Beijing International Studies University and Xi'an International Studies University). China's international qualification therefore needs to be improved.

The number of high-quality courses was relatively low, as evidenced by eight national-level quality courses and 34 provincial-level quality courses. Because the hotel management profession has high requirements for students' practical experience, establishing laboratories and offering laboratory courses are important indicators of the quality of professional education at colleges and universities. Roughly 80% of the universities had teaching laboratories and nearly 100% of third-tier schools had labs. However, laboratory equipment conditions and management standard need to be improved in hospitality programs. Only 43% of survey respondents found the laboratory equipment to be in satisfactory condition; about 49% were satisfied with laboratory management.

Within the past five years, professional teachers in hotel management had published 73 national-level planned textbooks. Among textbooks used by hotel

Table 3.4 Key performance indicators in university teaching assessment

	Total	Average	First-tier colleges	Second-tier colleges	Third-tier colleges
Major-related courses	518	10	13	10	7
Bilingual course ratio (%)	–	15.02	24.13	10.63	15
National-level elite courses	8	–	3	4	1
Provincial-level elite courses	34	–	12	19	3
Institution has lab (%)	–	80	68.75	82.93	100
Institutions has lab course (%)	–	64.67	60	67.42	53.33
Institution has on-site teaching (%)	–	59.32	56.25	57.5	100
National planned textbook published	73	1.55	19	51	3
Using English-language textbook (%)	–	7.83	14.29	4.04	10
Provincial award	90	2	3	2	1
Provincial and ministerial student awards	231	5	6	5	3
Compulsory internship requirement (%)	–	96.67	87.5	100	100
Internship period (months)	–	5.6	5.38	5.66	6
Graduate study for master's degree (%)	–	10.43	14.69	9.17	1
Graduate study for PhD (%)	–	5.15	19.45	1.5	0
Employment rate (%)	–	90.16	85.85	92.18	86.67
Hospitality industry employment (%)	–	59.07	50.17	61.69	69.33
Starting salary (RMB yuan/ month)	–	2,693	3,118	2,539	2,567

Note: Statistics for various institutions in the table are generally an average; however, due to the small amount of data related to the publication of national-level quality courses, provincial-level quality courses, and national planning textbooks, totals were used for these statistics.

management majors at various universities, 7.83% were English-language textbooks. Similar to bilingual teaching, the overall usage of English textbooks remains somewhat low.

In terms of teaching awards, participating universities had won 90 national and provincial teaching awards in the past five years. Hotel management students had won 231 national and provincial awards within the same period.

Hospitality programs in China focus on hotel industry integration. Among universities participating in the survey, only two had no fixed internship base. Regarding internship arrangements, schools generally stipulate a minimum duration for students. The shortest period was no less than two months and the longest was no less than one year. Most programs required a six-month internship.

The furthering graduate study rate is also an important indicator of teaching quality. According to the survey results, the average rate of postgraduate acceptance for hotel management majors was 10.4%, largely consistent with the rate for business management majors. Only 5% of applicants were admitted into PhD studies. The employment contract rate was about 90% on average.

However, the fit-in-industry proportion was only 59.07%. This figure represents a major concern for the education sector as well as for the industry. The starting salary for undergraduates in hotel management is approximately 2,000–3,000 yuan per month, much lower than that for business majors (4,000 yuan or more). This pay disparity is a critical issue that may prevent young talents from pursuing a hospitality career.

Academic research

Academic research is another core indicator that is of paramount importance for hospitality management programs. Table 3.5 lists the number of academic publications and social service-related studies reported in our survey.

Participating teachers published 751 papers within the past five years. Among them, 684 papers appeared in domestic academic journals (including Chinese Social Sciences Citation Index [CSSCI]) while 67 were featured in international journals (including SSCI). Additionally, the teachers had published a total of 129 monographs and 183 textbooks. They had been granted 338 national and

Table 3.5 National hotel management professional research, social service, and international cooperation

	Total	Average	First-tier schools		Second-tier schools		Third-tier schools	
			Total	Average	Total	Average	Total	Average
International journal publication	67	3	36	4	31	2	0	0
Domestic journal publication	684	18	250	25	426	16	8	4
Academic monograph	129	4	72	8	57	3	0	0
Textbook publication	183	5	48	5	129	5	6	2
National Natural Science Fund project	22	–	13	–	9	–	0	–
National Social Science Fund project	31	–	20	–	11	–	0	–
Ministry of Education Humanities and Social Sciences Fund project	48	–	21	–	27	–	3	–
Other provincial and ministerial projects	237	6	75	7	153	6	9	5
Commercial projects	346	13	255	24	81	6	10	5
Commercial project fund (10,000 yuan)	6,168	228.44	4,650	465	1,458	97.2	60	30
Total number of trainees at all levels (number of people)	32,594	741	16,064	1,236	15,230	544	1300	434
International cooperation project school (station)	9	–	1	–	8	–	0	–

provincial funded research projects: 22 were funded through the National Natural Science Foundation, 31 through the National Social Science Foundation, and 48 through the Humanities and Social Sciences Fund of the Ministry of Education.

Industry-related consulting projects are similarly important in hospitality research. Survey results revealed that 28 universities had undertaken 346 industry-related projects in the past five years, totaling 61.68 million yuan. On average, each university hosted 13 industry-funded projects totaling roughly 2,284,400 yuan.

International cooperation

International cooperation programs can be classified into two types: those accredited by China's Ministry of Education and those developed by a university. Among surveyed universities, nine were Sino-foreign cooperation projects accredited by the Ministry of Education of China and 22 had self-developed international programs.

Teachers' perceptions of the hotel management profession

Perceptions of the hotel management profession

Figure 3.4 depicts teachers' general attitudes regarding the future of the hotel industry profession. More than 70% of teachers were optimistic and only 5% were pessimistic.

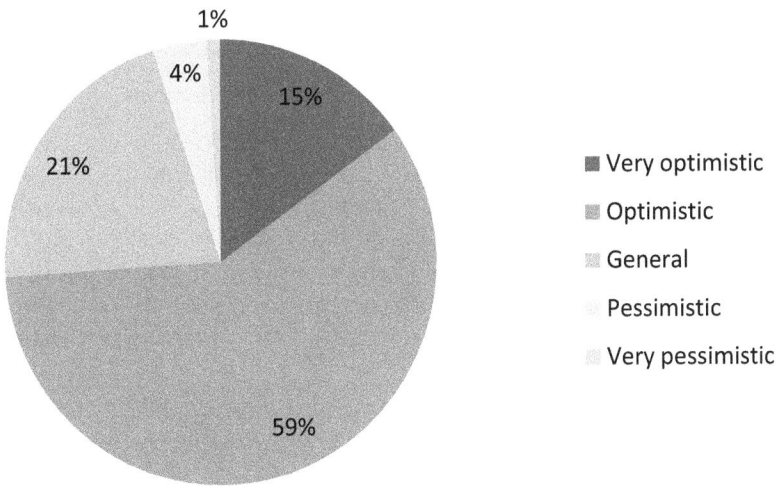

Figure 3.4 Teachers' perceptions of the hotel management profession.

Table 3.6 Perceptions of teachers' professional development prospects by school tier

	Very optimistic (%)	Optimistic (%)	Neutral (%)	Pessimistic (%)	Very pessimistic (%)
Overall	15	58.75	21.25	3.75	1.25
First-tier university	5.56	72.22	22.22	0	0
Second-tier university	15.52	55.17	22.41	5.17	1.73
Third-tier university	50	50	0	0	0

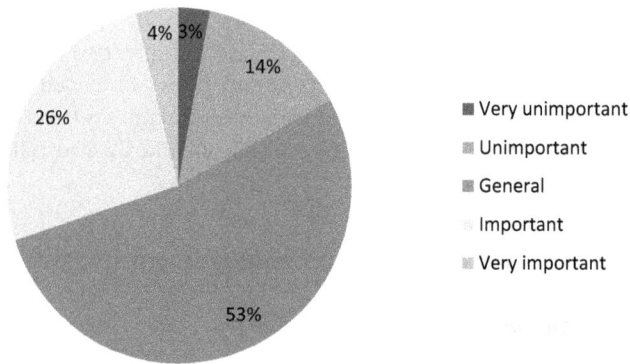

Figure 3.5 Teachers' perceptions of the importance of hotel management program.

Perceptions differed among universities: teachers at first-tier universities demonstrated less optimism than those at second- or third-tier universities (see Table 3.6).

Perceptions of the importance of hotel management programs

Contrary to perceptions of hotel management programs' development prospects, teachers' perceptions of these programs' importance at offering universities were relatively pessimistic. Figure 3.5 shows that only about 30% of surveyed teachers believed that the hotel management program plays an important role at their university.

Variation again emerged across universities: teachers at first-tier universities expressed less optimistic attitudes than those at second- and third-tier universities (see Table 3.7). Clearly, third-tier university teachers held much more positive perceptions than others.

Critical concerns from teachers

This survey also assessed teachers' main concerns. The most frequently mentioned issues included a lack of research resources, academic development prospects, and teaching quality.

Table 3.7 Teachers' perceptions of the importance of hotel management programs by school tier

	Very important (%)	Important (%)	Neutral (%)	Unimportant (%)	Very unimportant (%)
Overall	4	26	53	14	3
First-tier university	0	28	50	22	0
Second-tier university	2	28	55	12	3
Third-tier university	50	0	50	0	0

Compared with teachers in vocational schools, university teachers generally considered research performance to be a great challenge. Most teachers reported having difficulty to obtain necessary resources. Some teachers struggled with academic research due to a lack of familiarity with research in general. Because China does not yet boast any academic hospitality journals, teachers' publication performance and career development are correspondingly limited.

As a result, teachers in hospitality management have fewer promotion opportunities than their counterparts from tourism, other management-related fields, or geography. This disparity also reduces the individuals' motivation to focus on hospitality. Hotel management is a relatively new major, having first appeared in 2005. Its development therefore lags behind traditional majors such as tourism. Most teachers are quite interested in enhancing understanding of the hotel industry in general as well as of this discipline in particular. Many teachers are similarly eager to improve their professional qualities. Professional recognition among stakeholders is another crucial factor: teachers are often discouraged by the poor industry-fit employment rate, low start-up salary, and limited university recognition. Meanwhile, the COVID-19 pandemic has adversely affected the industry and dampened student admissions to related programs. This factor may negatively influence the major's future development.

Conclusions and recommendations

As indicated by the above analysis, hotel management education in China has developed rapidly and reached a high scale in terms of student enrollment as well as the number of teachers. As an independent major, it has established a curriculum system and achieved an independent academic field of study status. Yet, noteworthy problems persisted, including a lack of professional teachers, low research performance, an unappealing career path, and low discipline recognition. China recently launched the Double First-class Education Campaign. Several strategies can be implemented to establish this country as one featuring world-class hotel management education.

Integrating industry development needs

Rapid changes in the hotel industry have led to new requirements for the cultivation of hotel management professionals. Universities must keep up with industry needs and wants, especially those expressed by hotel groups. Schools must also stay abreast of technological advances to develop talent in areas such as big data, business analytics, and artificial intelligence applications. Doing so will boost industry-fit employment and high-end industry employment. Particular focus should be given to industry cooperation with respect to teaching, research, and internship opportunities.

Enhancing university recognition

Compared with the number of universities offering tourism management programs, the number of universities with hotel management programs is relatively low. Poor discipline recognition and a lack of appeal to teachers must be addressed. China's "New Liberal Arts" initiatives are expected to expand opportunities for such programs' innovation and development. Specific attention should be paid to integrating liberal arts as well as sciences into the hotel management program. It is expected that a multidisciplinary program may help to change the perception of hotel management from a low-value service profession to an academic field.

Enhancing internationalization

Although deglobalization may follow the COVID-19 pandemic to some extent, the world's hotel industry still needs globalized talent. It is hence essential to continue China's opening-up policy and to encourage more international cooperation. It is also recommended that more international programs be established between China and other countries. Additionally, online education should be encouraged at universities worldwide to build a closer but farther-reaching academic and educational community.

Acknowledgments

We would like to thank the China Tourism Education Association for providing information and research platforms. Thanks to the teachers who participated in the National Hotel Management Profession (Undergraduate) Teacher Training Course and to the WeChat group members who completed the online questionnaire; their survey contributions provided data-based evidence for our report. Thanks to Dr. Wang Yu, Dr. Li Pengbo, and the graduate students at BISU for their support in questionnaire design, questionnaire distribution, and research. Special thanks to our postgraduate students Song Xiaoxiao, Xu Kailun, and Yang Lulu for their translation and proofreading.

Appendix

Higher learning institutions in China offering hotel management undergraduate education (a total of 220 institutions)

Northeast	Anshan Normal College, Dalian University, Liaoning Normal University, Shenyang Institute of Technology
	Dongbei University of Finance and Economics, Eastern Liaoning University, Shenyang City University, Shenyang Normal University
	Dalian University of Finance and Economics, Jilin Business and Technology College, Jilin Huaqiao University of Foreign Languages
	Daqing Normal University, Qiqihar University, The Tourism College of Changchun University
	Harbin University, Harbin University of Commerce, Heilongjiang International University, Kunlun Tourism College
	Heilongjiang University of Finance and Economics
North China	Beijing International Studies University, Beijing Hospitality Institute, Beijing Jiaotong University, Haibin College
	Beijing Union University, Cangzhou Normal University, Hebei North University, Hebei University of Economics and Business
	Qing Gong College, North China University of Science and Technology, Jinzhong University
	Inner Mongolia University for The Nationalities, Shanxi University of Finance and Economics, Shanxi University
	Shanxi College of Applied Science and Technology, Shijiazhuang University, Taiyuan University, Tangshan College
	Tianjin University of Finance and Economics, Tianjin University of Finance and Economics, Pearl River College
	Tianjin Agricultural University, Tianjin University of Commerce, BinHai School of Foreign Affairs of Tianjin Foreign Studies University, Xinzhou Teachers University, China University of Labor Relations, Hulunbuir University
	Changzhi University, Handan College

Northwest	Gansu Normal University for Nationalities, Hexi University, Luoyang Institute of Science and Technology
	Qinghai Normal University, Shanxi Institute of International Trade & Commerce, Tianshui Normal University
	Xi'an International Studies University, Northwest Normal University, Xijing University
	Xinjiang University of Finance & Economics, Xinjiang University, Xi'an Innovation College of Yan'an University
	Xi'an Fanyi University, Xi'an Traffic Engineering Institute
Central China	Xinke College of Henan Institute of Science and Technology, Henan Normal University, Xinlian College of Henan Normal University
	Hubei University of Education, Hubei University of Economics, College of Law & Business of Hubei University of Economics
	Hubei University of Science and Technology, Hunan Institute of Technology, Hunan Women's University
	Hunan University of Commerce, Hunan International Economics University, Hunan Normal University
	Shuda College, Hunan Normal University, Hunan University of Arts and Science, Hunan Institute of Information Technology
	Huaihua University, Huanggang Normal College, Jishou University, Jianghan University
	College of Science and Technology of China, Three Gorges University, Songshan Shaolin Wushu College
	Wuchang University of Technology, Wuhan Qingchuan University, Wuhan Business University, Wuhan College, Xiangtan University
	Xuchang University, Changsha Normal University, Changsha University, Zhengzhou University
	Central South University of Forestry and Technology, Henan University of Science and Technology
	Luoyang Institute of Science and Technology, Wuhan Qingchuan University, Huanghuai University, Hubei University
	Hanjiang Normal University, Wuhan Donghu University, Hubei Business College, Wuhan Bioengineering Institute
	Xinxiang University, Xinyang University, Xinyang Normal University, Xinyang Agriculture and Forestry University
East China	Anhui Sanlian University, Anhui Foreign Language University, Changshu Institute of Technology
	Changzhou Institute of Technology, Chaohu University, Chizhou University, Chuzhou University
	Fujian Normal University, Fuyang Normal University, Gannan Normal University, Hefei Normal University
	Hefei University, Hohai University, Wentian College, Institute of Technology, East China Jiaotong University
	Huaqiao University, Huaibei Normal University, Huaiyin Normal University
	Huangshan University, University of Jinan, Jining University, Jiangsu University of Technology
	Jiangxi Science and Technology Normal University, Jiujiang University, Minjiang University
	Nanjing Tech University, Pujiang Institute, Nanjing Audit University, Jinshen College, Nantong University, Xinglin College
	Ningde Normal University, Qingdao Binhai University, Qingdao University, Xiamen University of Technology

Shandong Technology and Business University, Shandong Youth University of Political Science, Shanghai Polytechnic University, Sanda University, Shanghai Business School, Shanghai Normal University, Jiangsu Second Normal University

Anqing Normal University, West Anhui University, Economic and Technical College of Anhui Agricultural University

Xianda College of Economics and Humanities, Shanghai International Studies University

Applied Technology College of Soochow University, Tianping College of Suzhou University of Science and Technology

Wuyi University, Suzhou University, Yango University, Zhejiang Gongshang University

Zhijiang College of Zhejiang University of Technology, Zhejiang Yuexiu University of Foreign Languages, Jingdezhen University

Nanjing Xiaozhuang University, Qingdao Hengxing University of Science and Technology, Qingdao Huanghai University

Zaozhuang University

South China	Guangdong University of Finance & Economics, Guangdong University of Finance Guangdong, Peizheng College

Guangxi University of Finance and Economics, Lijiang College of Guangxi Normal University, Guangzhou College of Commerce

Guilin University of Aerospace Technology, Guilin University of Technology, Guilin Tourism University

Haikou College of Economics, Hainan University, Hainan Tropical Ocean University, Hainan Normal University

South China Normal University, Jinan University, Sanya University, Zhaoqing University, Sun Yat-sen University

Yulin Normal University, Baise University, Hezhou University, Guangxi University of Foreign Languages

Hanshan Normal University, Guangdong University of Foreign Studies, South China Business College

Southwest	Yinxing Hospitality Management College of CUIT, Chongqing Technology and Business University

Chongqing Universty of Science and Technology, Chongqing College of Humanities, Science & Technology

Chongqing Three Gorges University, Chongqing Normal University, Dali University, Guizhou University of Finance and Economics

College of Medicine, Guizhou Medical University, Honghe University, Kunming University of Science and Technology

Kunming University of Science and Technology, Oxbridge College, Kunming University, Leshan Normal University, Pu'er University, Qiannan Normal University for Nationalities, Qujing Normal University, Sichuan Institute of Industrial Technology

Sichuan University of Science & Engineering, Sichuan Tourism University, Sichuan Agricultural University

Sichuan Normal University, Chongqing Nanfang Translators College of SISU, Sichuan University of Arts and Science

Tongren University, Southwest Petroleum University, Yunnan University of Finance and Economics

Tourism and Culture College of Yunnan University, Zunyi Normal University, Yunnan Agricultural University

Guizhou Normal University, Guizhou University of Commerce, Moutai University, Southwest Petroleum University

Chongqing University of Education

Notes

1 Source: National Tourism Administration website, National Tourism Administration's statistical bulletin on culture and tourism in 2019. https://www.mct.gov.cn/whzx/ggtz/202006/t20200620_872735.htm.
2 The China Hotel Management Companies (Groups) Development Report 2018 issued by the China Tourists Hotel Association in 2019.
3 For a detailed list of the 220 hotel management schools, please see Table 3.1.
4 Due to the unique requirements of Chinese educational institutions, colleges with a high reputation and long history may be unlikely to launch doctoral programs because they are not top-tier schools and do not possess necessary certifications from China's Ministry of Education.

References

Bao, J., & Zhu, F. (2008). The problem and outlet of China's tourism undergraduate education shrinking: A reflection on the development status of tourism higher education in 30 Years. *Tourism Tribune*, 23(5), 13–17.

Bie, D., & Yi, M. (2014). The reality and policy response of China's higher education development. *Tsinghua University, Educational Research*, 35(1), 11–16.

Chen, F., & He, J. (2020). Emerging liberal arts education: Essence, connotation and construction ideas. *Journal of Hangzhou Normal University (Social Sciences Edition)*, 42(1), 7–11.

Chen, Y., & Dellea, D. (2015). Overview of Swiss hotel and tourism management education: Citations from the education experience of the Lausanne School of Hotel Management. *Tourism Tribune: Human Resources and Education Special Issue*, 30(S), 5–9.

Chen, X., & Zhan, L. (2009). Reflections on the development view of higher education in China. *Higher Education Research*, 30(8), 1–26.

Dai, B. (2005). On the teaching system and teaching methods of hotel management at the graduate level. *Tourism Tribune: Special Issues in Human Resources and Education*, 20(S), 183–186.

Dong, Z., Li, D., & Tan, Y. (2013). The dilemma and outlet of China's higher education management in the age of globalization. *Higher Education Research*, 34(10), 10–17.

Gu, H., Wang, J., & Zhang, X., & Zhang, W. (2003). World tourism education tour. *Tourism Tribune*, 18(S), 159–163.

Goodman, R. J., & Sprague, L. G. (1991). The future of hospitality education: Meeting the industry's needs. *Cornell Hotel and Restaurant Administration Quarterly*, 34(4), 90–95.

Hemmington, N. (1995). The attitudes of students to modular hospitality management programmes. *Education and Training*, 37(4), 32–37.

Lam, T., & Xiao, H. (2000). Challenges and constraints of hospitality and tourism education in China. *International Journal of Contemporary Hospitality Management*, 5(12), 291–295.

Li, F. L. (2020). Speeding up the construction of "new liberal arts" and actively leading the new era. *China Higher Education*, 2020(1), 45–47.

Liu, Z. (2014). From epitaxial development to connotative development: The value revolution of Chinese higher education in the era of transition. *Educational Research*, 35(9), 1–7.

Liu, Z., & Enz, C. (2006). The core resources and sustainable competitive advantage of Cornell hotel management school: Also on the enlightenment of China's tourism hospitality education. *Tourism Tribune: Human Resources and Education Special Issue*, 21(S), 29–35.

Penfold, P., Liu, W., & Ladkin, A. (2012). Developing hospitality education in China: A case study of Guilin Institute of Tourism. *Journal of China Tourism Research*, 8(1), 61–77.

Peng, Q. (1999). Exploration of higher vocational education in hotel management. *Tourism Tribune*, 14, 72–78.

Qin, Y. (2004). Analysis of several problems in hotel management teaching. *Journal of Beijing International Studies University*, 2004(1), 22–31.

Smith, S., & Ding, P. (2001). Origin, development, difficulties and pre-existence of tourism education in Australian universities. *Journal of Beijing International Studies University*, 2001(1), 24–31.

Tribe, J. (2001). Research paradigms and the tourism curriculum. *Journal of Travel Research*, 39(5), 442–448.

Wu, B., & Li, W. (2005). China tourism professional education development report. *Tourism Tribune: Human Resources and Education Special Issue*, 20(S), 9–15.

Xu, H., & Zhang, C. (2004). Comparative analysis and enlightenment of tourism education between China and foreign countries. *Tourism Tribune: Human Resources and Education Special Issue*, 19(S), 26–30.

Yu, C. (2008). Hotel higher education should receive more attention. *Tourism Tribune*, 23(3), 11–12.

Zou, Y., Chen, Y., & Qi, J. (2002). Analysis and countermeasure research on college students' loss of college students. *Business Economics and Management*, 134(12), 44–46.

4 Event management education in China

Qiuju Luo and Xiangru Qin

Introduction

Events have shown a great potential in attracting tourists with expenditure, accelerating the construction of infrastructure and extending the use of tourist-relevant facilities in host cities (Sou & McCartney, 2015; He et al., 2019). Targeting those advantages, economically advanced countries have raced to construct state-of-the-art convention facilities and host mega-events. China, as Asia's largest event market, boasted 13.73 million square meters (m^2) of venue capacity in 2018 (UFI, 2019) and ranked second to only the United States as the world's largest provider of exhibition space (He et al., 2019).

In the wake of the event industry's rapid development, higher education about event management has lagged behind considerably (Getz, 2002), especially in developing countries in East Asia (Sou & McCartney, 2015). Although institutions of higher education in Western countries began offering events-related majors in the 1970s, similar education in China has existed for only two decades. In 2001, Shanghai Art and Design Academy launched China's first tertiary vocational Program for event design, and undergraduate education in the discipline followed in 2002 when Beijing International Studies University began offering a program in event planning and management. By 2019, 125 universities and 235 vocational colleges were offering events-related Programs to students in China. Even so, a seeming paradox was occurring in China's event industry. Although companies repeatedly reported shortages of qualified workers, only 3.6% of graduates from events-related academic Programs had chosen jobs related to their majors (Sou & McCartney, 2015; Wang & Luo, 2018).

Higher education in event management, often practical in orientation, is pivotal to the industry (Getz, 2002). For this chapter, we first investigated the scale and scope of China's exhibition industry before the exploration into the higher education in event management. Next, our comparative investigation of host cities revealed a pyramid structure in the pattern of competition in the industry. Shanghai, Guangzhou and Beijing, as first-tier cities that host events, were the three cities that have hosted the most events, whereas Shenzhen and Tianjin led second-tier and third-tier cities, respectively. Last, an examination of event management education in both universities and vocational colleges in China

DOI: 10.4324/9781003004363-4

revealed that such education in the country has made substantial but unbalanced progress. Events-related graduate Programs continue to remain in their infancy, and the design of curricula at all levels of postsecondary education should be improved. Those and other findings provide insights into current problems and future trends in event management education in China.

China's event industry and pattern of competition

Following Luo (2018), we used four indicators to investigate the current development of China's exhibition industry: number of exhibitions, average exhibition area, total area of exhibition venues and number of the global trade association for the exhibitions industry (UFI) members and UFI-approved events. With reference to statistics from 2015 (Luo, 2018), our work reveals six features of China's exhibition industry:

- China is the largest market for trade fairs in the Asia-Pacific region, although its development remains unbalanced.
- In 2019, 28 cities in China sold more than 1 million m^2 of exhibition area. The cities that sold more than 10,000 m^2 grew from one (i.e. Shanghai) to three (i.e. Shanghai, Guangzhou and Beijing) from 2015 to 2019.
- Small- and medium-sized exhibitions continue to dominate the market share, even as numerous large-scale events and mega-events have been developed.
- China has a sufficient amount of exhibition space, with 29 cities in 2018 capable of providing an indoor exhibition area exceeding 0.1 million m^2.
- The number of both UFI members and UFI-approved events has risen, while the four cities (i.e. Shanghai, Beijing, Guangzhou and Shenzhen) have retained their above-average advantage of scale.
- A pyramid structure shapes the pattern of competition in China's exhibition industry and implies that its development has remained uneven.

Number of exhibitions and amount of exhibition space sold

The number of exhibitions held in cities is a vital index of health in the event industry. As the largest market for trade fairs in the Asia-Pacific region, China hosts an exceptionally large number of exhibitions in various cities across the country. According to the China Council for the Promotion of International Trade (CCPIT), China hosted 3,547 exhibitions in 2019, and Shanghai in particular held more than 400 exhibitions and organised 545 fairs that year or 15.4% of all domestic exhibitions (CCPIT, 2020). Apart from Shanghai, five other cities—Beijing, Guangzhou, Qingdao, Zhengzhou and Shenzhen—each organised more than 100 exhibitions in 2019, and the exhibitions organised by those six cities accounted for 43.56% of China's total that year (CCPIT, 2020), for a relatively high concentration ratio nationwide.

Table 4.1 List of cities in China selling more than 1 million m^2 of exhibition space in 2019 (unit = million m^2)

Rank	City	2015	2019	Rank	City	2015	2019
1	Shanghai	15.12	26.12	15	Wuhan	3.07	2.43
2	Guangzhou	8.62	15.02	16	Nanjing	2.25	2.39
3	Beijing	5.20	11.02	17	Changchun	1.71	2.19
4	Shenzhen	2.78	5.10	18	Xiamen	1.91	1.88
5	Qingdao	2.95	4.26	19	Dongguan	3.40	1.82
6	Zhengzhou	2.30	4.24	20	Linyi	–	1.76
7	Chengdu	3.10	3.95	21	Shijiazhuang	–	1.75
8	Hangzhou	2.65	3.71	22	Urumqi	–	1.71
9	Chongqing	7.02	3.35	23	Dalian	1.18	1.61
10	Shenyang	3.10	3.00	24	Ningbo	1.52	1.60
11	Tianjin	3.46	2.68	25	Kunming	–	1.59
12	Jinan	1.79	2.66	26	Hefei	1.86	1.43
13	Changsha	1.25	2.54	27	Nanning	–	1.13
14	Xi'an	2.60	2.48	28	Harbin	1.97	1.13

Sources: Luo (2018); China Council for the Promotion of International Trade (2019).

The total space sold at exhibitions, related to the number of exhibitions and the total area of an exhibition, comprehensively captures the state of development in the exhibition industry. In 2019, total exhibition space sold at trade fairs amounted to 130.48 million m^2, and 28 cities sold more than 1 million m^2 in exhibition space (see Table 4.1). In 2019, compared with 2015, when only Shanghai sold more than 10 million m^2 (Luo, 2018), the exhibition space in each of the three top-tier cities (i.e. Shanghai, Guangzhou and Beijing) exceeded more than 10 million m^2. In addition, the area of exhibitions held in the four leading cities (i.e. Shanghai, Guangzhou, Beijing and Shenzhen) totalled 57.29 million m^2 (45.35%), which was also larger than that in 2015 (i.e. 35.96 million m^2, 33.16%). Those results also indicate an increased high-level concentration ratio in China's exhibition industry.

Average size of exhibition

The average scale of exhibition is another crucial index to measure the current status of event industry. In 2019, China's event industry was dominated by exhibitions ranging from 0.1 to 1 million m^2 in area (42%), followed by those with less than 0.1 million m^2 (40%). Compared with the market share in 2015, small-scale exhibitions that sold less than 0.1 million m^2 in space increased by 7.3% in 2019. Growth, albeit by only 0.3%, was also experienced among exhibitions with 1–5 million m^2 of space. Small- and medium-sized exhibitions have thus continued to dominate the market share of China's exhibition industry.

The average area of an exhibition is the ratio of the exhibition's area to the number of exhibitions, which reflects the average size of exhibitions held in one city. Generally speaking, large-scale exhibitions have strong industrial influence and attract more attendees. In 2019, 23 cities in China held more than

Table 4.2 Chinese cities ranked by average area of exhibition space in 2019 (unit = 10,000 m^2)

Rank	City	2015	2019	Rank	City	2015	2019
1	Shenzhen	3.12	3.14	13	Changsha	–	1.23
2	Changchun	1.39	1.94	14	Zhengzhou	0.90	1.18
3	Shanghai	2.02	1.92	15	Guiyang	–	1.16
4	Beijing	1.25	1.85	16	Ningbo	1.12	1.13
5	Kunming	–	1.82	17	Wuhan	0.91	1.11
6	Chongqing	0.94	1.82	18	Hangzhou	0.89	1.10
7	Guangzhou	1.79	1.62	19	Shenyang	0.89	1.07
8	Tianjin	1.33	1.50	20	Dalian	–	1.06
9	Jinan	1.21	1.46	21	Xiamen	0.99	1.04
10	Chengdu	1.59	1.46	22	Hefei	1.02	1.03
11	Qingdao	1.47	1.40	23	Nanjing	1.00	0.95
12	Xi'an	1.44	1.23				

Sources: Luo (2018); China Council for the Promotion of International Trade (2019).

100 exhibitions per year and sold space amounting to more than 10,000 m^2 (see Table 4.2). Relative to figures from 2015, 13 of those 23 cities experienced an increase in their average exhibition area. Those results indicate that China's exhibition industry cultivates a certain number of large-scale events and mega-events with significant industrial influence.

Area of exhibition venues

The number and area of exhibition venues, as essential infrastructure, reflect the development of a city's exhibition industry. Referring to official statistics, Table 4.3 lists the 27 cities in China with the total indoor exhibition area exceeding 0.1 million m^2 in 2018. Shanghai topped the list with 977,000 m^2 of space, followed by Guangzhou with 492,400 m^2 and Kunming with 389,800 m^2. The total exhibition area in China grew from 27.80 million m^2 in 2018 to 40.26 million m^2 in 2019, an increase of 44.81%.

The statistics reveal that large pavilions are clustered in large cities, including Beijing, Shanghai, Guangzhou and Shenzhen, whose ample exhibition facilities with large indoor areas afford them a competitive edge in the exhibition industry. In June 2020, the largest indoor exhibition facility in China was Shenzhen International Convention and Exhibition Centre, with an area of 0.5 million m^2, 0.4 million m^2 of which was built in the first phase. Shanghai National Convention and Exhibition Centre ranked second, with an area of 0.4 million m^2, followed by exhibition space in Guangzhou with an area of 0.45 million m^2. Although less in exhibition space than Shenzhen and Shanghai, Guangzhou has sufficient facilities and infrastructure for large-scale exhibitions. Beijing is also equipped with many exhibition venues, including the New China International Exhibition Centre with 0.11 million m^2 of indoor exhibition space, while Wuhan, Qingdao, Nanjing and Chengdu all host exhibition venues with more than

Table 4.3 Chinese cities with a total indoor exhibition area of more than 100,000 m² in 2018 (unit = 0.1 million m²)

Rank	City	2018	Rank	City	2018
1	Shanghai	977.0	15	Linyi	176.6
2	Guangzhou	492.4	16	Xi'an	170.0
3	Kunming	389.8	17	Nanjing	161.1
4	Chengdu	325.0	18	Nanchang	156.0
5	Hangzhou	307.6	19	Shenyang	129.6
6	Chongqing	305.2	20	Yiwu	126.4
7	Qingdao	295.0	21	Zibo	123.0
8	Beijing	289.6	22	Zhongshan	117.5
9	Changchun	227.9	23	Wuxi	112.5
10	Suzhou	223.0	24	Zhuhai	106.2
11	Wuhan	220.4	25	Shenzhen	105.0
12	Binzhou	213.0	26	Tianjin	101.0
13	Foshan	196.0	27	Xiamen	100.0
14	Wenzhou	194.0			

Source: China Convention, Exhibition, Event Society (2019).

Table 4.4 Geographic distribution of UFI members and UFI-approved events across China in 2016 and 2020

	Number of UFI members		Number of UFI-approved events	
	2016	2020	2016	2020
Beijing	29	28	30	23
Shanghai	24	26	18	21
Guangzhou	10	17	7	11
Shenzhen	11	14	11	12

Source: Luo (2018); UFI (2020).

0.1 million m² in area. Overall, cities in China offer sufficient indoor space for exhibitions.

UFI members and UFI-approved events

Corporate membership in the UFI, which ranks among the most important global associations in the exhibition industry, reflects the essential status of an enterprise or institution in the industry. In January 2020, the UFI had 172 members in China (see Table 4.4). The four leading cities in terms of membership—Beijing, Shanghai, Guangzhou and Shenzhen—accounted for 37.79% of all members. Beijing ranked first with 28 members (16.28%), followed by Shanghai with 26 (15.12%).

Approval from the UFI is an essential credential for high-quality exhibitions. Any party interested in hosting an exhibition accepted as a UFI-approved event needs to meet the UFI's auditing requirements concerning the size, history and internationalisation of applicants. In 2020, China had 143 domestic exhibitions

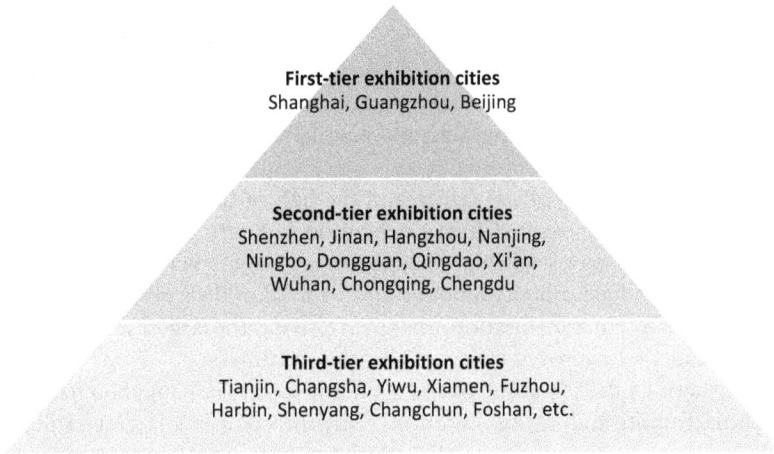

Figure 4.1 The pattern of competition in China's exhibition industry.
Source: Adapted from Luo (2018).

approved by the UFI, most of which were held in Beijing, Shanghai, Jinan, Shenzhen and Guangzhou. Beijing ranked highest with 23 (16.08%) UFI-approved events, followed by Shanghai with 21, Jinan with 16, Shenzhen with 12 and Guangzhou with 11. The number of certified exhibitions in the top five cities accounted for 58.04% of the total. Compared with statistics from 2016, figures from 2020 indicate a rising trend in three cities: Shanghai, Guangzhou and Shenzhen. Although Beijing retains a compelling above-average advantage of scale, the number of UFI members and UFI-approved events in Beijing has slightly decreased since 2016.

In sum, a particular gap exists in China's exhibition industry, primarily manifested by the industry's concentration in Shanghai, Beijing, Guangzhou and Shenzhen. These urban centres have a higher degree of branding and offer considerably more large-scale exhibitions and UFI-approved events than other cities in China.

Patterns of competition in China's exhibition industry

Following Jin, Bauer and Weber (2010) and Luo (2018), we highlight the pyramid structure in the pattern of competition in China's exhibition industry (see Figure 4.1), the development of which has remained uneven among cities in China (Zhou et al., 2017).

China's first-tier cities for exhibitions

China's first-tier cities for exhibitions—Shanghai, Guangzhou and Beijing—are equipped with a large number of exhibition venues with sufficient indoor capacity and thus host the bulk of international exhibitions that convene in China (Jin et al., 2010). Among them, Shanghai leads the pack in terms of number of exhibitions, average exhibition area, and total area of exhibition venues. Shanghai is followed by Guangzhou and Beijing. All either provincial capitals or developed urban centres with global reputations (Zhou et al., 2017), those three cities are located in eastern, southern and northern China, respectively, and lead development in their surrounding areas.

As an international financial centre and the hub of the Yangtze River Delta, Shanghai ranks foremost in China's exhibition industry. In terms of number of exhibitions and area available, Shanghai has absolute superiority as well as offers sufficient venue capacity amounting to a total of 0.49 million m². With a high degree of internationalisation, Shanghai also has the largest number of UFI members and UFI-approved events of all cities in China.

As southern China's largest city and economic centre, Guangzhou has hosted the China Import and Export (Canton) Fair, the country's largest exhibition, every spring and autumn since 1957. Without a doubt, the Canton Fair has played a driving role in the development of Guangzhou's exhibition industry. In 2012, Guangzhou surpassed Beijing in exhibition space sold and thus ranked second only to Shanghai, and in 2020, it boasted an exhibition area totalling more than 0.1 million m². Although the UFI memberships and UFI-approved events in Guangzhou are fewer than in Beijing, Guangzhou has clearly risen to the status of China's second-leading city for exhibitions.

Last, as China's political and economic centre, Beijing has a high international reputation, ranking second only to Shanghai in the number of UFI memberships and UFI-approved events. However, fewer exhibitions are held in Beijing than in Guangzhou, and Beijing has often had to transfer small- and medium-sized exhibitions to other locales due to political events. Adverse environmental problems, including traffic congestion, also contribute to the shrinking development of the exhibition industry in the city.

China's second-tier cities for exhibitions

Albeit with relatively large numbers of exhibitions and sufficient area for them, China's second-tier cities typically lack the indoor capacity and degree of internationalisation in the first-tier cities. According to Jin, Bauer and Weber (2010), the second-tier cities are typically categorised as either provincial capitals (e.g. Jinan, Hangzhou, Nanjing, Xi'an, Wuhan and Chengdu) or economically developed cities (e.g. Shenzhen, Ningbo, Dongguan and Qingdao).

As the leading second-tier city, Shenzhen occupies a top position in the number and size of exhibitions and even ranks first in average exhibition area due to the Shenzhen Convention and Exhibition Centre's massive capacity of 105,000 m². Beyond that, Shenzhen's exhibition industry is highly internationalised. In 2019, the city ranked third in China in terms of the number of UFI members and UFI-approved events, behind Shanghai and Beijing. Shenzhen has also introduced and cultivated several influential brand exhibitions both at home and abroad.

Jinan, a major city in the Circum-Bohai-Sea region, is a vital hub for equipment manufacturing in China. Relying upon its developed industrial base, Jinan hosts many brand exhibitions, including the China (Jinan) International Solar Energy Utilization Conference and Exhibition. In 2020, Jinan surpassed Shenzhen and Guangzhou and became the third-largest city in China in terms of hosting UFI-approved events.

Xi'an and Wuhan, the respective capitals of Shanxi and Hubei Provinces, lead the development of the exhibition industry in central China, whereas Chengdu and Chongqing play essential roles in western China's exhibition economy. Although the number and area of exhibitions held in Chengdu and Chongqing are among the highest in China, their degree of internationalisation remains rather low.

On the heels of the top-ranked Shanghai, the cities of Hangzhou, Nanjing and Ningbo have achieved rapid development in the exhibition industry. By the number and area of exhibitions, they lead the pack; however, their degree of internationalisation needs to be improved.

Last, the city of Dongguan, located in the Pearl River Delta and the Guangdong–Hong Kong–Macao Greater Bay Area, is a major manufacturing centre. Leveraging its local industrial system, Dongguan has sought to develop its exhibition industry, has cultivated many brand exhibitions (e.g. for furniture) and occupies a substantial market share when it comes to exhibitions in China.

China's third-tier cities for exhibitions

Compared with China's first- and second-tier cities, its third-tier cities for exhibitions—for instance, Tianjin, Changsha, Yiwu, Xiamen, Fuzhou, Harbin, Shenyang, Changchun and Foshan—host the least number of exhibitions and offer the least amount of exhibition space. Nevertheless, they have held some influential exhibitions and demonstrate remarkable market potential.

Tianjin, a major city in the Beijing–Tianjin–Hebei region and Circum-Bohai-Sea region, has a solid foundation of heavy industries, high flows of port traffic and a developed logistics industry. The number and scale of Tianjin's exhibitions place it at the fore of China's third-tier cities for exhibitions. With an advantage in the market share of medium-sized exhibitions (i.e. 0.3–0.5 million m^2), the city's capacity for exhibitions is relatively large. Although the Meijiang Convention and Exhibition Centre is the only pavilion available in the city, the National Convention and Exhibition Centre (Tianjin) is currently under construction and slated for completion in September 2021.

Other third-tier cities, including Changsha, Yiwu and Xiamen, are affected by more developed cities for exhibitions in their surrounding areas. Although many such cities do not lack large, modern conference or convention facilities, they do lack international recognition, professional event companies and skilled labour (Zhou et al., 2017).

As those findings imply, a noticeable gap exists between China's large cities and their smaller counterparts when it comes to hosting exhibitions. A critical deficiency in second-tier and third-tier cities is the real or potential absence of a professionally qualified workforce (Zhou et al., 2017).

Profile of event management education in China

The sustained popularity of events worldwide justifies a great demand for professional talent and, in turn, should promote higher education in event management, typically called *event management education*. In short, event management education aims to cultivate well-educated, qualified staff for the event industry (Zeng & Yang, 2011). In China, such education has developed into a comprehensive system of hierarchies (see Figure 4.2), with two categories of schools that provide events-related programs:

1   Vocational schools at the secondary level, which offer vocational programs lasting two to three years and associate programs lasting five years and that award a college diploma.
2   Institutions of higher education, which offer four-year bachelor's programs, two- to three-year master's programs and three- to four-year doctoral programs.

Scale of development

Event management education in China has a history spanning only two decades. In 2001, the first vocational program for event design was established in Shanghai Art & Design Academy, followed by an undergraduate program at Beijing International Studies University in 2002, which marked the official beginning of higher education in event management in China.

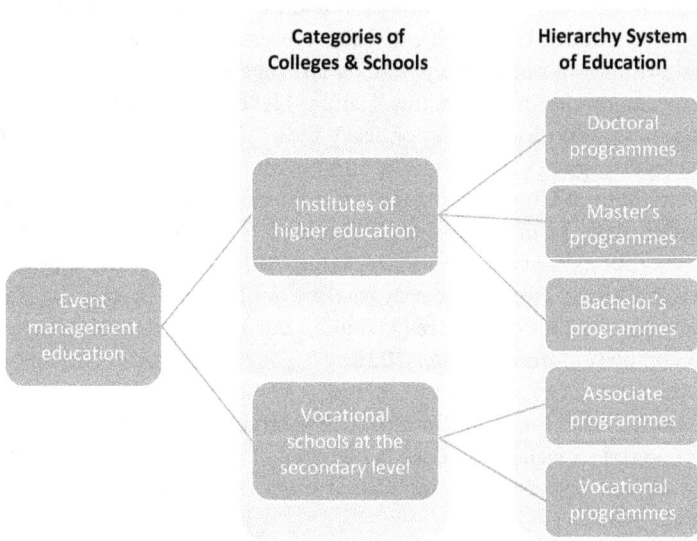

Figure 4.2 China's system of event management education.
Source: Adapted from Yin and Meng (2018).

Since 2015, event management education in China has expanded rapidly (see Figure 4.3). Statistics reveal that whereas only 276 colleges and schools offered events-related programs in 2013, that number surged to 443 in 2019. Meanwhile, the number of vocational schools with such programs at the secondary level, after fluctuating only slightly from 203 to 192, rose to 265 in 2019 and continued to exceed the number of institutions of higher education with such offerings.

From 2004 to 2018, the number of students in events-related programs grew as well (see Figure 4.4). Students in such Programs at vocational schools, totalling 4,798 in 2008, exceeded 10,000 by 2014 but declined slightly in 2015.

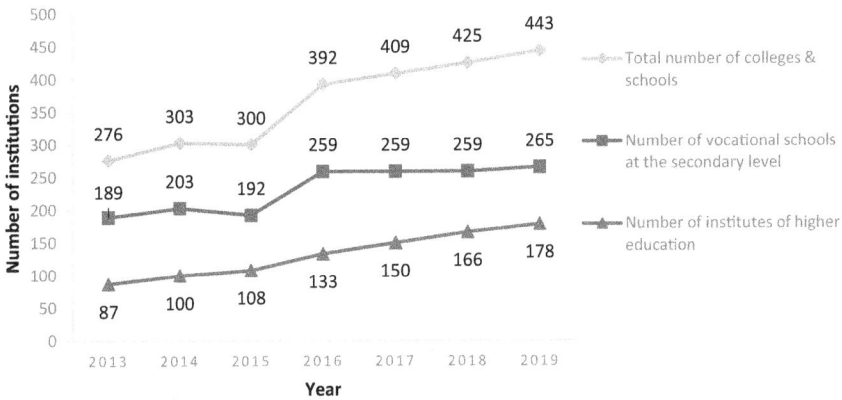

Figure 4.3 Number of colleges and vocational schools with event management programs in China, 2013–2019.

Source: Statistic is extracted from database of the Ministry of Education of China (2013–2019).

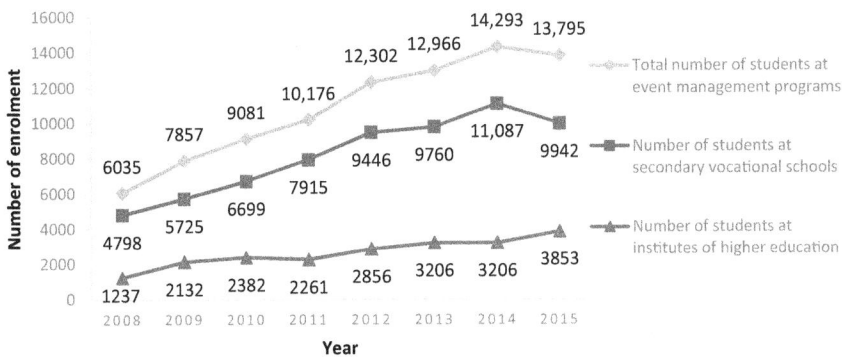

Figure 4.4 Number of students in programs for event management at colleges and vocational schools in China, 2008–2015.

Source: Zhejiang Convention & Exhibition Society (2015).

In contrast, the number of students in such programs at institutions of higher education reached 3,853 in 2015, showing a twofold growth since 2008. Nevertheless, the number of students at vocational schools continued to exceed that at institutions of higher education.

China's event management education system

Event management is commonly characterised as a discipline area that 'focuses on a particular phenomenon and draws relevant theory and knowledge from other fields and disciplines' (Getz, 2002, p. 15). Because event management is not an independent discipline, most colleges and institutions situate it in their departments of tourism and hospitality.

Vocational and associate programs

According to the data collected in 2019 from China's Ministry of Education, events-related majors were offered at 265 tertiary-level vocational colleges in 28 provinces, municipalities and autonomous regions of Mainland China. Vocational and associate programs in the field are both offered at secondary-level vocational schools; students in associate programs usually finish two to three years of vocational coursework before pursuing a diploma by spending two additional years in college (Yin & Meng, 2018). In 2019, only five vocational schools provided associate programs in event management.

China's secondary-level vocational schools offer five majors associated with event management: event planning and management (n = 162, 61.13%), exhibition art design (n = 36, 13.58%), digital exhibition technology (n = 33, 12.45%), wedding planning and management (n = 21, 7.92%) and cloth design and exhibition (n =13, 4.91%). Of the mentioned 265 colleges, 33 (12.45%) offer at least two types of those majors.

The geographic distribution of events-related programs at vocational colleges appears in Table 4.5. By tier of city, the number of colleges with such programs increases as the number of cities in the tier increases. The first tier, consisting of only three cities, has the least total number of vocational colleges with each individual major. Although the second tier includes 11 cities, vocational colleges in those cities do not outnumber those in their first-tier counterparts. Because cities not specified in the first or second tier are categorised in the third tier, this tier has the most vocational colleges with such programs, distributed across more than 30 cities.

From tier to tier, the distribution of events-related majors is unbalanced (see Table 4.5). Above all, the dominant program is event planning and management. The sole exception is Wuhan, where programs in event planning and management and digital exhibition technology are one and the same. First-tier cities offer at least three events-related programs and thus show relatively comprehensive development. Nevertheless, a program for cloth design and exhibition does not exist in any first-tier cities. By contrast, six of the 11 second-tier cities (i.e.

Table 4.5 Events-related majors established by vocational colleges in China by tier of city

Tier	City	P & M	Art	Wedding	Digital	Cloth D & E
First	**Subtotal**	**36**	**6**	**4**	**3**	**0**
	Shanghai	11	3	1	0	0
	Guangzhou	17	2	2	2	0
	Beijing	8	1	1	1	0
Second	**Subtotal**	**37**	**8**	**6**	**7**	**1**
	Shenzhen	0	0	0	0	0
	Jinan	4	1	1	0	0
	Hangzhou	6	2	0	0	0
	Nanjing	2	2	1	1	0
	Ningbo	1	0	0	0	1
	Dongguan	1	0	0	0	0
	Qingdao	4	0	0	0	0
	Xi'an	3	0	0	0	0
	Wuhan	4	1	2	4	0
	Chongqing	5	0	1	1	0
	Chengdu	7	2	1	1	0
Third[a]	**Subtotal**	**90**	**22**	**11**	**23**	**12**

Source: Statistic is extracted from database of the Ministry of Education of China (2020).
Note: P & M = exhibition planning and management; art = exhibition art design; wedding = wedding planning and management; digital = digital exhibition technology; cloth D & E = cloth design and exhibition.
a   Third-tier encompasses more than 30 cities; therefore, the data of third-tier cities are not listed in detail due to page limitation.

Shenzhen, Hangzhou, Ningbo, Dongguan, Qingdao and Xi'an) have no more than two vocational colleges with programs in event management. Altogether, the uneven distribution of events-related programs could fail to cultivate certain types of workers and thereby exacerbate shortages in human resources in local events industries. That trend worsens in third-tier cities, most of which have established only vocational programs in event planning and management, outnumbering other events-related programs.

Undergraduate programs

Universities in China currently offer two types of majors related to events: exhibition economics and management, and exhibition art and technology. Overall, 125 universities specialise in exhibition economics and management, far more than those specialising in exhibition art and design. University programs for event management education are primarily grounded in the fields of economics, management and marketing.

China's Ministry of Education first approved the major in exhibition economics and management in 2004. Only two universities—Shanghai Normal University and Shanghai University of International Business and Economics—were authorised to offer the major at the time. Beijing International Studies University was formally approved to establish the major in 2006, after which increasingly more colleges and universities applied to establish event programs. In 2012, the Ministry of Education adjusted its *Catalogue of Disciplines and Majors for Undergraduate Programs at General Institutions of Higher Education*, and thus officially changed the exhibition economics and management program from a pilot to a formal program. Altogether, from 2004 to 2019, the Ministry of Education approved 125 institutions or universities to establish undergraduate programs for exhibition economics and management.

By contrast, exhibition art and technology was first established at Shanghai University in 2005. From 2005 to 2012, the Ministry of Education approved 16 colleges and universities to offer undergraduate programs in this specialisation. In 2012, when such programs became integrated into Schools of Art, Science and Technology, they were relabelled as art and technology or the exhibition of art and technology. From 2004 to 2019, the Ministry of Education recorded 53 institutions or universities specialising in exhibition art and technology.

In terms of geographic distribution, the number of institutions of higher education with those programs is roughly higher in tiers with more cities (see Table 4.6). Thus, the third tier has the most programs in exhibition economics and management. However, the lion's share of programs in exhibition art and

Table 4.6 Number of institutions of higher education offering events-related programs in cities in China by tier

Tier	City	Exhibition economics & management	Exhibition art & technology
First	**Subtotal**	**25**	**10**
	Shanghai	10	5
	Guangzhou	9	2
	Beijing	6	3
Second	**Subtotal**	**34**	**24**
	Shenzhen	0	0
	Jinan	4	2
	Hangzhou	5	4
	Nanjing	1	6
	Ningbo	1	0
	Dongguan	0	0
	Qingdao	1	0
	Xi'an	4	3
	Wuhan	8	2
	Chongqing	4	2
	Chengdu	6	5
Third[a]	**Subtotal**	**67**	**10**

Source: Statistic is extracted from database of the Ministry of Education of China (2020).
a   Third-tier encompasses more than 30 cities; therefore, the data of third-tier cities are not listed in detail due to page limitation.

technology are in second-tier cities. Meanwhile, first-tier cities have the fewest institutions of higher education with programs offering either major.

From tier to tier, the distribution of events-related majors at institutions of higher education remains uneven. First-tier cities offer both programs, although those awarding degrees in exhibition economics and management dominate. Only 18.18% of the second-tier cities have established programs in exhibition economics and management, while Shenzhen and Dongguan do not yet offer any events-related programs in their institutions. In the third tier, 33 of 42 cities have only one events-related program. In all, the development of events-related programs in China's institutions of higher education continues to be imbalanced.

Postgraduate programs in event management

Two types of postgraduate programs in event management are provided at China's institutions of higher education: a Master of Science degree, which is awarded to full-time postgraduates performing academic research for two or three years, and a Master of Tourism Administration (MTA), which is granted to part-time postgraduates who complete a corresponding program in two to five years (Yin & Meng, 2018).

On the one hand, the Master of Science program aims to foster graduates with a scientific mindset and an ability to conduct research. According to Yang (2019), 31 universities offer such events-related postgraduate programs; most of them (n = 16, 51.61%) have established event management as a subfield of tourism management, whereas six (19.35%) offer programs in event management itself. Other universities situate events-related postgraduate programs into their Departments of Business Administration (12.9%), Communications (6.45%) or Public Administration (3.23%).

On the other hand, a few universities offer MTA programs, which are generally designed to deliver applied knowledge from cases and nurture professional leaders in the event industry. Eight universities (88.89%) offer an events-related MTA program under tourism management, while one university regards event management as a sub-discipline of communications (Yang, 2019).

In terms of geographic distribution, universities offering events-related postgraduate programs are primarily located in first-tier cities (38.71%), followed by third-tier cities (29.03%). In the second tier, such programs appear in only three cities: Chengdu, Qingdao and Nanjing. Such imbalance implies a discrepancy between the rapid development of the event industry and the supply of experts in the field.

Doctoral programs in event management

The scale of doctoral programs in event management remains relatively small. In 2015, only three doctoral programs focused on event management (Yang, 2016), and two of the corresponding universities confer doctoral degrees in tourism management. The exception is Shanghai University, which offers an advertising and events program in the Department of Journalism and Communications (Yang, 2016).

Clearly, the development of events-related doctoral programs in China remains in its infancy. The three mentioned programs are all located in first-tier cities—Guangzhou (*n* = 2) and Shanghai (*n* = 1). In 2017, Sun Yat-sen University in Guangzhou established the first postdoctoral program in event management.

Curriculum design

The heart of educational endeavour is curriculum, which should be designed to meet the expectations of students, industry actors, researchers and society at large (Airey et al., 2015). A well-defined curriculum should also be able to produce well-rounded, identifiable talent (Oktadiana & Chon, 2017). However, vocational and university bachelor's programs may offer equivalent and/or indistinct curricula, thereby creating uncertainty on the job market for graduates (Gu et al., 2007).

To compare event management curricula between vocational and university programs, we selected a vocational college and an institution of higher education, both located in Guangzhou. Coursework was obtained from the transcripts of graduates, and the courses were coded into eight groups: strategic planning and management, operations and implementation, marketing, communications, MICE technology, economics and business studies, mathematics and research methodology and law, psychology and general studies (see Table 4.7).

Among the results, some courses were included in both vocational and university bachelor's programs. In strategic planning and management, for instance, five courses in both programs were comparable: Management Principles, On-Site Event Management, Development of the Event Industry, Project Management and Human Resources Management. That combination was also applied in other courses, including Tourism and Event Marketing, English, Space Design in Exhibitions and Laws and Ethics.

The courses in both vocational and university programs aligned with those highlighted by industry professionals (Zeng & Yang, 2011; Huang et al., 2017). Zeng and Yang (2011), for instance, collected 69 surveys from event practitioners and identified 12 courses based on their five top-ranked courses, all of which stressed a focus on event planning, law and policy, marketing, project management and English. By comparison, Huang et al. (2017), using the Delphi method, proposed ten top professional priorities for event personnel, the highest of which was a good capacity for communication and expression. This finding implies alignment between the combination of courses and the expectations of the industry. Thus, although both vocational and university bachelor's programs were roughly in line with the development of the event industry, caution is needed in future event management curriculum due to the fast-growing nature of the event industry (Zeng & Yang, 2011).

Second, a discrepancy surfaced between the vocational and university bachelor's programs. On the one hand, practical activities for students in vocational programs should be emphasised (Oktadiana & Chon, 2017), with four courses

Table 4.7 Courses in event management curricula in vocational and university bachelor's programs in China

	Vocational bachelor's programs	Combined vocational and university bachelor's programs	University bachelor's programs
Strategic planning and management	Mega-Event Planning and Management Event Customer Relationship Management Introduction to International Events	Management Principles On-Site Event Management Development of the Event Industry Project Management Human Resources Management	Introduction to Conferences and Exhibitions Meetings Planning and Management Exhibition Planning and Organisation Festival Management
Operations and implementation	Event Etiquette and Service Conference Operation Management Simulated Event Practice Exhibition Construction		Service Internship Specialised Practice
Marketing	Exhibition Network Marketing Telemarketing Technology Event Advertising Brand Planning	Tourism and Event Marketing	Consumer Behaviour Market Case Study
Communications	Foreign Trade Correspondence Business Negotiation and Eloquence Training	English	Cross-Cultural Communication
MICE technology	Introduction to Information Technology Graphics and 3D Design Event Logistics	Space Design in Exhibitions	
Law, psychology and general studies	Psychological Services	Law and Ethics Current Situations and Policy Employment Guidance Sports and Health	
Economics and business studies	Introduction to International Trade		Microeconomics Macroeconomics Accounting Financial Management Industry Economics Business Statistics
Mathematics and research methodology			Research Methodology Advanced Mathematics Probability and Statistics

focusing on operations and implementation. This suggestion was also applied in the fields of marketing, communications and MICE technology. Functional communication skills (e.g. business negotiation and eloquence training) were also conferred to students in the vocational programs.

On the other hand, the in-depth study of theory should be stressed for students in event management programs at universities (Oktadiana & Chon, 2017). In such programs, the curriculum designated for university students contained more general and theoretical knowledge in marketing, economics and business. For example, the programs offered in-depth knowledge about consumer behaviour, micro- and macroeconomics, accounting, financial management and industry economics. Beyond that, university bachelor's students learned more about mathematics and research methods, which were entirely absent in the vocational programs.

In sum, similarities and distinctions characterise vocational and university education in event management in China's institutions of higher education. Whereas vocational education focuses on developing the technical skills required for a specific job, university education emphasises transferable skills and knowledge (Oktadiana & Chon, 2017).

Conclusion and directions for future development

In China, the event industry has experienced rapid yet imbalanced development. Cities, including those in the second and third tiers, have sufficient convention facilities but suffer from a shortage of professionally skilled labour (Sou & McCartney, 2015; Zhou et al., 2017). Needless to say, the progress of event management education has lagged behind that of the industry. To promote the future development of event management education, we have the following three recommendations.

Promoting the balanced development of events-related programs

Event management education at vocational colleges and institutions of higher education has made substantial but uneven progress. In vocational and associate programs, event planning and management remains the dominant major, while others, including exhibition and art design and digital exhibition technology, are emerging in programs that nevertheless need to be developed. A similar trend is occurring in undergraduate programs, in which the primary major is also exhibition economics and management. Events-related education should not only have management fundamentals, but also draw from various other disciplines and fields to access diverse cultural, environmental, legal, psychological and political perspectives on events and the event industry (Getz, 2002).

Given the pyramid structure of the pattern of competition for exhibitions among China's cities, the uneven development of events-related programs is worse in second- and third-tier cities. Some cities (e.g. Shenzhen) offer only one program in event management at vocational or other institutions of higher education. The shortage of professionally qualified workers thus hampers the development of the event industry there and in other second- and third-tier cities. Indeed, 57.5% of business managers reported that those cities lack professional event companies (Zhou et al., 2017). Therefore, promoting the balanced development of events-related Programs, especially in China's second- and third-tier cities, needs to be prioritised.

Expanding postgraduate education in event management

A discrepancy has arisen between the event industry's rapid development and the supply of experts in the field. In China, events-related graduate programs remain in their infancy, and although event management education promotes a strong orientation towards industry and employment, graduate-level studies need to involve investigations of theoretical, methodological and ethical issues in industry practices (Getz, 2002). Talented professionals with critical thinking skills and capacities to cope with complex problems and uncertainty are highly valuable for the fast-growing event industry (Oktadiana & Chon, 2017).

At the same time, faculty members in event management education have diverse educational backgrounds. Most have transferred from other relevant disciplines—economics, management, geography and marketing, among others—and may lack a thorough understanding of events and industry practices. For a solution, they need to upgrade their knowledge by interacting with industry practitioners (Yin & Meng, 2018). Meanwhile, postgraduate and doctoral programs should be enhanced to cultivate a new generation with systematic professional education in event management.

Improving curriculum design

Event management education has been categorised as either vocational or academic in focus (Oktadiana & Chon, 2017). Academic education stresses theories and trends in the MICE industry, although industry professionals have expressed a less positive perception of the research practices taught (Zeng & Yang, 2011). Others have blamed a heavy outflow of talent in the event industry and the mismatch between theory and practical skills learned in academic education (Gu et al., 2007). In response, Sou and McCartney (2015) have advised MICE undergraduate programs balance professional and functional knowledge and sharpen vocational skills and internships. Even so, Wang and Luo (2018) have empirically revealed that the industry's high turnover rate is partly attributed to the contradiction between short MICE career paths and individuals' career development needs. In that light, the curriculum for event management should not be solely driven by a pragmatic design (Getz, 2002).

Instead of balancing theoretical and practical approaches, Mayaka and Akama (2015) have emphasised the importance of relating curricula to context and outcomes. Amid the rapid development of the event industry, the curriculum for event management should be designed to meet the fast-changing needs of industry, researchers, students and society at large. An ideally balanced curriculum for the discipline incorporates both management fundamentals and theoretical knowledge while cultivating good citizens and event professionals (Getz, 2015). Thus, event management education in China should foreground 'the experience of learning, the deep, intimate connections between knowledge and daily life, and the capacity to develop critical, mindful and reflexive practice' (Airey et al., 2015, p. 13).

References

Airey, D., Dredge, D., & Gross, M.J. (2015). Tourism, hospitality and events education in an age of change. In D. Dredge, D. Airey & M.J. Gross (Eds.), *The Routledge handbook of tourism and hospitality education* (pp. 3–14). New York: Routledge.

China Convention, Exhibition, and Event Society. (2019). *Report of 2018 Chinese exhibitions*, 8 April 2019. Available from: http://www.cces2006.org/index.php/home/index/detail/id/12252/ [23 June 2020].

China Council for the Promotion of International Trade. (2020). *Annual report on Chinese exhibition economy 2019*. Available from: http://aaa.ccpit.org/Category7/Asset/2020/Jan/10/onlineeditimages/file71578657177531.pdf [28 June 2020].

Getz, D. (2002). Event studies and event management: On becoming an academic discipline. *Journal of Hospitality and Tourism Management*, 9(1), 12–23.

Getz, D. (2015). Event higher education: Management, tourism and studies. In D. Dredge, D. Airey & M.J. Gross (Eds.), *The Routledge handbook of tourism and hospitality education* (pp. 476–491). New York: Routledge.

Gu, H.M., Kavanaugh, R.R., & Yu, C. (2007). Empirical studies of tourism education in China. *Journal of Teaching in Travel & Tourism*, 7(1), 3–24.

He, H.W., Li, C.X., Lin, Z.B., & Liang, S. (2019). Creating a high-performance exhibitor team: A temporary-organization perspective. *International Journal of Hospitality Management*, 81, 21–29.

Huang, H.C., Hou, C.I., Wu, S.L., Chang, I.Y., & Lin, C.H. (2017). Study on work capacity for MICE personnel. *International Journal of Organizational Innovation*, 9(4), 303–313.

Jin, X., Bauer, T., & Weber, K. (2010). China's second-tier cities as exhibition destinations. *International Journal of Contemporary Hospitality Management*, 22(4), 552–571.

Luo, Q.J. (2018). Profile and trend of the exhibition industry in China. In J.L. Zhao (Ed.), *The hospitality and tourism industry in China: New growth, trends, and developments* (pp. 175–240). Oakville: Apple Academic Press.

Mayaka, M.A., & Akama, J.S. (2015). Challenges for the tourism, hospitality and events higher education curricula in sub-Saharan Africa: The case of Kenya. In D. Dredge, D. Airey, & M.J. Gross (Eds.), *The Routledge handbook of tourism and hospitality education* (pp. 3–14). New York: Routledge.

Oktadiana, H., & Chon, K. (2017). Vocational versus academic debate on undergraduate education in hospitality and tourism: The case of Indonesia. *Journal of Hospitality & Tourism Education*, 29(1), 13–24.

Sou, K.I.S., & McCartney, G. (2015). An assessment of the human resources challenges of Macao's meeting, incentive, convention, and exhibition (MICE) industry. *Journal of Human Resources in Hospitality & Tourism*, 14(3), 244–266.

UFI (2016). *Search UFI members 2016.* Available from: https://www.ufi.org/membership/ufi-members/search/ [20 June 2016].

UFI (2019). *UFI releases new trade fair industry in Asia report*, August 2020. Available from: http://www.ufi.org/wp-content/uploads/2019/08/Report-Trade-Fair-Industry-in-Asia-Media.pdf/ [28 June 2020].

UFI (2020). *Search UFI members 2020.* Available from: https://www.ufi.org/membership/ufi-members/search/ [28 June 2020].

Wang, M.S., & Luo, Q.J. (2018). Exploring the MICE industry career path. *International Journal of Contemporary Hospitality Management*, 30(5), 2308–2326.

Yang, Q. (2016). *National conference and exhibition undergraduate schools (2002–2015)*, 14 January. Available from: https://mp.weixin.qq.com/s/hoL-0Re8IE5hRIKQxSoaNA/ [30 June 2020].

Yang, Q. (2019). *Overview of 2019 national convention and exhibition vocational colleges*, 19 November. Available from: https://m.expomanager.cn/45/80/p668674035f4942/ [30 June 2020].

Yin, Z.X., & Meng, F. (2018). Tourism higher education in China: Profile and issues. In J.L. Zhao (Ed), *The hospitality and tourism industry in China: New growth, trends, and developments* (pp. 241–261). Oakville: Apple Academic Press.

Zeng, X.H., & Yang, J. (2011). Industry perceptions of the event management curriculum in Shanghai. *Journal of Convention & Event Tourism*, 12(3), 232–239.

Zhejiang Convention & Exhibition Society. (2015). *China Meeting, Incentive, Conference and Exhibition Education Development Report 2015*. Available from: https://www.qufair.com/news/2015/12/23/1075.shtml [28 June 2020].

Zhou, C., Qiao, G.Q., & Ryan, C. (2017). How might Chinese medium sized cities improve competitive advantage in the event tourism market? *Event Management*, 21(1), 109–118.

5 Tourism vocational education

Kunxin Wang, Jian Hu,
Angzhi Gao and Zhichao Yang

Introduction

Tourism vocational education in China started at the same time with the Reform and Opening up and developed along with the industry of tourism. Over the past 40 years, starting from scratch, China's tourism vocational education witnessed its growth from being small and weak to being large, strong and prosperous. Now tourism vocational education in China is the world's largest tourism vocational education system, with the largest number of tourism colleges, majors, teachers and students.

This chapter has five sections, namely an overview of tourism vocational education, a review of the development of tourism vocational education in China over the past 40 years, the personnel training and social services in vocational tourism education, the innovations and characteristics of tourism vocational education and the challenges and opportunities faced by vocational tourism education.

Overview of tourism vocational education

Among all types of education, vocational education is closely connected with society and economy. The contemporary Chinese vocational education began in the early 20th century, and developed slowly in the subsequent 60 years. At the end of the 1970s, when China began the reform and opening up, vocational education began to draw attention and gain rapid development. After nearly 40 years of vigorous development, vocational education has become an important part of China's education.

With the rapid development of tourism industry, China's tourism vocational education has been constantly developing and expanding. It has gradually established a distinctive education system, covering all levels of secondary vocational education, higher vocational education, applied undergraduate education and so on. As an important way to train professional personnel for China's tourism industry, tourism vocational education has prepared a large number of qualified skillful professional personnel for the tourism industry.

DOI: 10.4324/9781003004363-5

Understanding tourism vocational education

Vocational education is an educational ideology as well as an independent academic field. According to the *International Standard Classification of Education* issued by the UN Educational, Scientific and Cultural Organization (UNESCO) in 2011, vocational and technical education is divided into three levels: primary (2B), secondary (3B) and higher (5B) (UNESCO, 2011). Primary and secondary vocational and technical education is defined as skill-oriented education and higher vocational and technical education is defined as profession-oriented education. It is an international common practice to grant the trainee a qualification certificate on some certain skills rather than a diploma of certain degree after his/her completion of vocational education. In China's education system, vocational education is divided into secondary vocational education and higher vocational education. In order to ensure that graduates can enter the industry smoothly, they are required to obtain corresponding skill certificates before graduation.

Tourism Vocational Education aims to train professional personnel for tourism industry. In a broad sense, tourism vocational education refers to the education that equips the trainees with professional knowledge, skills and professional ethics needed for certain occupations based on the requirements for the employees in the tourism industry. In a narrow sense, tourism vocational education refers to the process of the systematic and professional training carried out by the educators in a purposeful, planned and organized way, based on the requirements of tourism industry for professional personnel.

The purpose of tourism vocational education is to train practice-oriented personnel and high-quality workers with professional knowledge and skills. Compared with general education, tourism vocational education focuses more on the training of practical skills and practical working ability.

Tourism vocational education is an educational system with distinct practical characteristics. Vocational education is adaptable to the changes of the market and the demands of the society through its diversified management, flexible training system and emphasis on students' practical abilities. Therefore, tourism vocational education programs, especially those programs in higher vocational colleges, promote flexibility, diversity and applicability, with closer contacts with enterprises to enhance graduates' employability.

In comparison with vocational education in other industries, tourism vocational education demonstrates the following characteristics: first, it serves the tourism industry; second, it focuses on professional services along the industrial chain of tourism in the six fields of "food, accommodation, transportation, sightseeing, shopping and entertainment"; third, it values the close connection to the industry and enterprises; and fourth, it stresses the integration of theory into practice. As a labor-intensive industry with most of its employees directly serving the tourists, tourism emphasizes practical capability and sympathy. Therefore, tourism vocational education must highlight the significance of practice and apply relevant theories into industry practice. In addition, tourism vocational teachers should have the required professional work attitude and ethics.

Mission and value of tourism vocational education

Tourism vocational education takes training professional personnel for tourism industry as its highest mission. Serving a certain industry or occupation in a specific stage of their developments, vocational education needs to design the specifications of the training to meet personnel needs of the industries and positions. Therefore, the quality standards for "the training process" and the "product output" need to be formulated and evaluated by the industry. With the changes of market economy and the innovations in science and technology, the requirements of enterprises for professional personnel are constantly changing. Vocational education graduates are required to meet the demands for the development of enterprises in the industry in a timely manner so as to constantly provide power for the sustainable development of enterprises in the industry. Through joint research, the colleges of vocational tourism education and industrial enterprises determine the specialty setting, define the training target and design the training mode and curriculum system, so that the specialty construction and training will fully reflect the trend of the development, characteristics and requirements of production, operation and technology for the enterprises in the tourism industry. On one hand, the industry and the enterprises find their source of professional personnel from colleges and universities; on the other hand, the schools proactively adapt to the requirements of the industrial development and train various professional personnel for the industry.

The demand from industry development for professional personnel is the driving force to promote reforms in educating and teaching, which is mainly reflected in the following aspects: First, they will promote reforms in programs. Based on the reality and the trend of industrial development, vocational colleges would adjust the settings of programs, such as wellness tourism, establish the information release and dynamic adjustment warning mechanism of settings of programs and promote the connection between the settings of different programs. Second, they will promote reforms in curriculum. The structure and content of curriculum will be designed according to the professional qualification standards, the actual requirements of the industry and enterprises. The employers will directly participate in curriculum design, evaluation and the introduction of international advanced courses, which will quicken the response of vocational education to technological progress and enhance its ability in serving small- and medium-sized innovative enterprises. Third, they will promote reforms in education. The reforms in the contents of teaching will truly reflect the industrial development, technological innovation and social development; the reforms in the process of teaching will reflect the process of actual production and service; the reforms in the methods of education will arouse the enthusiasm of the students towards learning through solving the real problems in the process of practice, practical training, internship and research-based learning. Fourth, they will promote reforms in classroom teaching. Regular adjustment and improvement systems of the curriculum will be established to keep the curriculum and teaching materials updated, so as to meet the needs of the industry and society. The curriculum will be optimized with reduced required courses and increased elective courses for more options of the students; the teaching methods will be optimized by promoting small-sized classes, streamed classes and bilingual education to improve the quality of classroom teaching.

The construction of training base for vocational education needs to be strengthened in order to train professional personnel to satisfy the demands of the markets. During the construction of the training base, the school shall conduct full investigation to achieve seamless connection with industry enterprises, elaborate the construction scheme, which will provide the real workplace scenario for the students to learn. The training base shall keep pace with the industry and go ahead of the industry moderately so as to prepare the students with new technology, new process and new methods. At the same time with the construction of training base, virtual simulation training base can be applied and developed in training with the utilization of digital teaching resources.

It is one of the missions of tourism vocational education to serve tourism with knowledge and skills. In order to further promote the tourism vocational education to serve the local tourism economy and promote the development of tourism industry, the tourism higher vocational education at present mainly serves the industry in the following aspects: First, setting up relevant research institutions and tourism think tanks to serve tourism development. For example, the research base for tourism vocational education in Shandong College of Tourism and Hospitality and the research base for tourism standardization in Tourism College of Zhejiang, established by China Tourism Academy, have greatly promoted the participation of tourism vocational colleges in tourism industry research; second, school-government and school-local authority cooperation mechanisms have been established and improved. For example, Tourism College of Zhejiang has established a Development Council with 20 Industry-Institution collaboration workstations based on the cooperation mechanisms between institution, government agencies and local businesses, respectively; third, the tourism vocational colleges are playing bigger roles in tourism economy. Many colleges and universities have undertaken a large amount of personnel training for local enterprises, scientific research projects, to serve the development of local tourism as a booster of local tourism economy.

In recent years, the reform in the system and mechanism of cooperation between enterprises and universities or colleges became the "top concern" of higher vocational education as well as the inevitable requirement to ensure the sound development of vocational education. The current cooperation mechanism between universities or colleges and enterprise follows the modes below: The first is the widely promoted mode of order-oriented, tailor-made training, which has become the trump card of tourism vocational colleges. The second is the improved level of cooperation between enterprises and universities or colleges. The cooperation has grown from merely the affairs of colleges or universities into the concern of the alliance of enterprises and colleges or universities. Relying on large tourism enterprises and tourism colleges, vocational education groups founded by provinces and cities have formed alliances between enterprises and colleges or universities. The third is the ever-deepening cooperation between enterprises and colleges or universities in school's perspective. **It is also an indispensable mission to promote the all-round development of young people, alleviate the poverty and backwardness through education.** One of the significant tasks of vocational education is to help the poverty alleviation through education:

> To vigorously develop vocational education and training, with the focus on improving the basic education and technical skills of the registered low-income people, so as to comprehensively improve the over-all abilities of the people from the poverty-stricken areas in going out of poverty through employment or starting business.

This is the ultimate goal of vocational education in promoting poverty alleviation put forward in *The Plan for Poverty Alleviation through Education during the "13th Five-Year" Plan* jointly issued by six departments, including the Ministry of Education (Ministry of Education, 2016).

Classification of tourism vocational education

Tourism vocational education can be classified into two levels: secondary vocation education and higher vocational education. Specialized study fields include tourism management, hotel management, travel agency service, exhibition management, scenic spot service, culinary art, catering management and so on (Figure 5.1).

Secondary vocational education mainly takes the forms of technical secondary school of tourism, vocational high school of tourism and vocational technical schools of tourism, which are designed to train workers, junior- and middle-level managers for the tourism industry. These schools have existed for a relatively long time. They are many in number and have concentrated majors and large

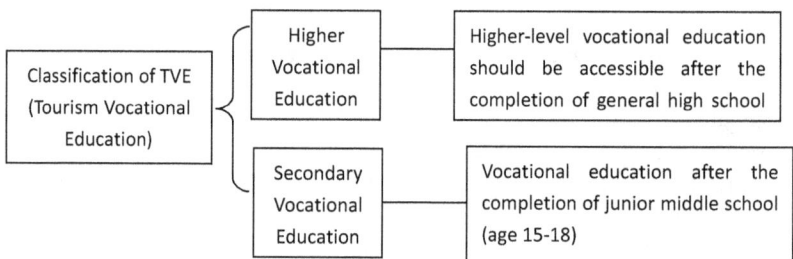

Figure 5.1 Classification of tourism vocational education.

enrollment of students. In terms of the current situation of the tourism industry, a large number of primary- and middle-level managers, service staff and workers are needed. The most suitable personnel of this level and type are the graduates of secondary tourism vocational schools. They are equipped with both theoretical knowledge and practical abilities. Moreover, the cost of training a secondary vocational school student is far lower than that of training a college student.

Higher tourism vocational education falls into the forms of tourism vocational colleges, adult colleges and senior technical colleges. These higher education institutes take the responsibilities of training workers, grassroots management personnel and skilled technical personnel for the tourism industry. They are large in scale with many schools, a variety of programs and a large number of students (Table 5.1).

In China, the scale of higher vocational education of tourism is larger than that of secondary vocational education of tourism in terms of the number of students or teachers, number of programs or the coverage of programs and the impact. So it is particularly important to develop higher vocational education of tourism.

The tourism vocational education system in China is reflected in three aspects: level, teaching and teacher. After 40 years of development, these three aspects have been basically formed and constantly improved in tourism vocational education. **In terms of level**, with regards to academic education, higher vocational education and secondary vocational education have been established at different levels, including high, middle and low, while in the field of the non-degree earning education, a variety of training modes, such as on-the-job training, adult education and further overseas studies have been formed. Such a multidimensional and multilevel education system has been relatively complete in the structure of training and basically met the needs of industry development. **As for teaching**, most of the tourism colleges have formed the teaching system of teaching materials, experiments, classroom teaching, practical teaching, curriculum construction and program construction. The whole teaching system is linked with different levels of teaching in various featured forms, showing distinctive characteristics. **In terms of teacher**, the important role of "dual-profession" teachers with practical experience has been highlighted in tourism education. Teaching staff of tourism colleges have basically formed a framework with academic teachers in school, professional teachers from the industry, teachers of theoretical education and teachers for practical training (Wang, 2018).

According to the *Statistics on National Tourism Education and Training 2017* released by the Ministry of Culture and Tourism (April 17, 2018), there are 1,086 general colleges and universities with higher vocational tourism management programs (mainly including seven majors of Tourism Management, Tour Guide, Travel Agency Operation and Management, Scenic Area Development and Management, Hotel Management, Leisure Service and Management and Exhibition Planning and Management) and 947 secondary vocational schools with tourism-related programs (including five majors: Operation and Management of High Star-rated Hotels, Tourism Service and Management, Foreign Languages for Tourism, Tour Guide Service, Exhibition Service and

Table 5.1 Classification of tourism vocational education programs

Level	General category	Name of program	Code of program
Four-Year Undergraduate Program of Higher Vocational Education	3401 Tourism	Tourism Management	340101
		Hotel Management	340102
		Tourism Planning and Design	340103
	3402 Catering	Culinary Arts and Food Service Management	340201
Three-Year Program of Higher Vocational Education	5401 Tourism	Tourism Management	540101
		Tour Guide	540102
		Travel Agency Operation and Management	540103
		Tailor-Made Travel Management and Service	540104
		Educational Tourism Management and Service	540105
		Hotel Management and Digital Operation	540106
		B&B Management and Operation	540107
		Wine Culture and Marketing	540108
		Tea Ceremony and Tea Culture	540109
		Smart Scenic Sites Development and Management	540110
		Smart Travel Technology Application	540111
		Convention and Exhibition Planning and Management	540112
		Leisure Service and Management	540113
	5402 Catering	Smart Catering Management	540201
		Culinary Arts and Nutrition	540202
		Chinese and Western Pastry Technology	540203
		Western Food Technology	540204
		Diet and Nutrition	540205
	4102 Forestry	Forest Ecotourism and Wellness Tourism	410210
	5003 Water Transportation	International Cruise Service	500304
	5004 Air Transportation	Flight Attendant	500405
	5702 Language	Tourism English	570203
		Tourism Japanese	570207
Secondary Vocational Education	7401 Tourism	Travel Service and Management	740101
		Tour Guiding Service	740102
		Health Tourism and Leisure Service	740103
		Deluxe Hotels Operation and Management	740104
		Tea Ceremony and Marketing	740105
		Convention and Exhibition Service and Management	740106
	7402 Catering	Chinese Culinary Art	740201
		Western Culinary Art	740202
		Chinese and Western Pastry	740203
	7702 Language	Tourism Foreign Languages	770209

Sources: *The Catalogue of Programs of Vocational Education (2021)*, Ministry of Education of the People's Republic of China.

Management). It is estimated that there are about 340,000 students studying in higher vocational education and 230,000 students in secondary vocational education (Ministry of Culture and Tourism, 2018). China has become the country with the largest-scale tourism vocational education in the world.

Review of tourism vocational education during the past four decades

The development of tourism vocational education in China has roughly experienced the following three stages:

The initial stage (1978–1995)

After the restoration of order in the political field and economic reform and opening up, vocational education in China has regained momentum in its development since 1978. Two national working conferences on vocational education were held in 1986 and 1991, respectively. In 1991, the State Council's *Decision to Vigorously Develop Vocational and Technical Education* made clear provisions on the nature, status, role, direction, tasks and measures of vocational and technical education, and reiterated the issue about the establishment of primary, secondary and higher vocational and technical education systems. The *Outline for Reform and Development of Education in China*, issued in 1993, put forward a clear direction and goals and tasks for further reform and development of vocational education. Accordingly, a series of regulations and basic principles for the management system, school running system, investment system, teaching, staff development and evaluation standards in vocational education are formulated. All of these showed the efforts and determination of the government in promoting the development of vocational education. In the same period, China's tourism industry began to shift from political reception to economic benefit-oriented development, with an increase in the number of international tourist arrivals and foreign exchange earnings. The development and transformation of the tourism industry called urgently for a large number of high-quality tourism professionals. Under the two premises of the national development of vocational education and tourism industry's earnest demands for professional tourism workers, China's tourism vocational education has started and gradually developed in line with the development of China's tourism industry.

Tourism vocational education emerged to meet the demands from the development of tourism, accelerating the training of tourism professionals, on-the-job personnel and high-level management personnel. In 1978, Jiangsu Provincial Tourism School (now Nanjing Institute of Tourism and Hospitality) was founded as the first tourism secondary school in China and the first tourism college, Shanghai Institute of Tourism, was established the following year. The establishment of these two schools marked the beginning of tourism vocational education in China.

In particular, two adult colleges are worth mentioning here. They are namely China Tourism Management Institute and Jinling Hotel Management Institute. Adult education, which lays no restrictions on age or gender, refers to the education different from the formal full-time education. Through the adult education system, people regarded as adults in the society can obtain needed skills, knowledge and technical and professional qualifications. There are four main forms of adult continuing education: self-study examination of higher education (self-study examination), correspondence education (distant education), adult college entrance examination (off-job, amateur and correspondence) and open universities (modern distant open education by the former radio and television universities). The adult higher education in China belongs to the category of vocational education.

China Tourism Management Institute, as an economic and technical assistance project jointly invested by the Chinese government and the UN Development Program, is the first tourism adult higher education institute in China. It was officially approved by the Ministry of Education of the People's Republic of China to be set up in Tianjin in 1987. It is one of the large-scale higher education institutions specialized in tourism education in China. On May 14, 2010, with the approval of the Ministry of Education and the National Tourism Administration, China Tourism Management Institute was merged into the Tourism Department of Nankai University (currently the College of Tourism and Service Management).

Jinling Hotel Management Institute, as a college for training professionals for tourism and hotel services management under the auspice of the Tourism Bureau of Jiangsu Province, was established in 1989 with the approval of the State Education Commission (now the Ministry of Education) and Jiangsu Provincial Government. In 2001, with the approval of Jiangsu Provincial Government, it was merged into Nanjing Tourism School and later upgraded to Nanjing Institute of Tourism and Hospitality in 2008.

Founded in the late 1980s, both schools trained a group of managers for the tourism industry in China in the early development times and played important roles in the development of tourism education and tourism industry in the last century.

The China National Tourism Administration attached great importance to the development of secondary tourism schools in all provinces not only in terms of macro management guidance, but also in funding support. Many regional universities and universities under central government ministries also established tourism programs, schools and colleges. Among them are the secondary tourism vocational schools like Sichuan Institute of Tourism, Hubei Provincial Tourism School, Zhejiang Provincial Tourism School, Guangdong Provincial Tourism School, Shandong Provincial Tourism School, Jiangxi Provincial Tourism School, Shaanxi Provincial Tourism School, Zhengzhou Tourism School, Taiyuan Tourism School and Shenyang Tourism School, Qingdao Tourism School, tertiary tourism vocational colleges like Sichuan Cuisine College (now Sichuan Tourism University) and Guilin Tourism College (now Guilin Tourism University). In terms of vocational training, as early as in the 1950s, there were a limited number of tourism vocational training programs in China. In the 1980s, these vocational training programs gradually transformed into adult diploma education and training. After decades of efforts, secondary tourism vocational education has substantially developed and become an important part of tourism education in China.

In 1995, according to incomplete statistics, there were altogether 622 tourism colleges and universities in China, including 138 institutions of higher learning, 17 secondary tourism schools and 467 vocational high schools and technical schools. There are nearly 140,000 enrolled students, including more than 20,000 in colleges and universities, 8,800 in secondary schools and 110,000 in vocational colleges and technical schools. There are more than 20 programs in these colleges and universities, covering four categories of tour guide, management, technology and teaching. In terms of adult vocational education, by 1995, two tourism adult colleges had been established in China with about 700 on campus students; the self-taught college entrance examination by Central TV and Radio University, staff universities, night universities and some local education departments had offered some tourism programs and awarded college or undergraduate degree diploma. Colleges and universities with tourism departments or programs, secondary tourism schools and more than ten specialized training institutions in various regions provided a wide range of training for tourism practitioners.

This stage marks an important period in the development of secondary tourism vocational education.

Development stage (1996–2005)

In 1996, the National People's Congress passed and promulgated the *Vocational Education Law of the People's Republic of China*. Since then, vocational education in China has embarked on the road of standardized development in accordance with the law. In 1998, the National People's Congress passed and promulgated the *Higher Education Law of the People's Republic of China*, which made it clear that higher vocational education should be vigorously developed to cultivate a large number of professionals with necessary theoretical knowledge and strong practical ability. By this time, many domestic vocational colleges, independent adult colleges and some technical colleges had gradually shifted into vocational and technical colleges (or vocational colleges) through reform, reorganization and restructuring, which accelerated the development of higher vocational education in China. Meanwhile, in 1996, the number of tourists in China's tourism industry exceeded 50 million for the first time and the tourism foreign exchange revenue exceeded 10 billion US dollars, which brought China into the top ten countries in international tourism receipts. The national legislation for vocational education and the status of China as a large tourism country pushed tourism vocational education into a period of large-scale development. In this period, there were two new changes in the development.

First, tourism vocational colleges experienced the change from extensive development to intensive development. Taking higher vocational colleges as an example, the tourism higher vocational education in China in this period enjoyed a considerable scale and a relatively good development. At the end of the 20th century, the state required all colleges and universities and adult colleges to train "high-level applied technical professionals" as higher vocational and technical colleges had been doing, and the education implemented by these colleges and universities was collectively referred to as "higher vocational and technical education". The scale of tourism vocational education in this period could basically meet the development of local tourism. Generally speaking, tourism higher vocational education is relatively developed in the regions with a developed tourism industry, meeting the local demands on trained tourism professionals. Statistics in 2005 showed that there were more than 50 colleges and universities offering tourism programs in each of the provincial-level regions of Beijing, Hebei, Liaoning, Heilongjiang, Jiangsu, Anhui, Fujian, Shandong, Henan, Hunan, Guangdong, Chongqing and Sichuan, with Hunan topping all of them with 104 schools. Basically, these provinces were the regions with relatively developed tourism economy, among which five provinces, including Jiangsu, Zhejiang, Fujian, Shandong and Liaoning, were the national top ten in terms of tourism foreign exchange in that year.

Second, the status of tourism discipline has been further recognized. In June 2004, China's first catalog of programs in higher vocational colleges was officially promulgated. The first catalog of programs in higher vocational education determined the division principle primarily based on industry and trade with the disciplinary classification taken into account and adopted the three-level structure of "program cluster—program category—program", which provided the basis for the construction and reform of the programs in higher vocational education.

In this period, with the rapid development of tourism and the rush of secondary vocational schools' upgrading, some secondary vocational schools had been upgraded to tourism vocational colleges. Among them, Zhejiang Tourism School took the lead to upgrade to Tourism College of Zhejiang in 2000. In 2002, Qingdao Vocational and Technical College of Hotel Management and Jiangxi Tourism and Commerce Vocational College were founded, followed by Changsha Commerce & Tourism College in 2003 and Zhengzhou Tourism College, Taiyuan Tourism College, Shanxi Vocational College of Tourism and Shandong College of Tourism and Hospitality in 2004.

Independent tourism vocational colleges had been successively established. As of June 2020, there were 26 independent tourism vocational colleges in China, 11 of which were upgraded or established as vocational colleges between 2000 and 2005. This period saw the rapid development of higher vocational tourism education. The details are shown in Table 5.2.

Quality consolidation stage (2006–2020)

From 2006 to 2015, the Ministry of Education and the Ministry of Finance launched the "National Model (Key) Higher Vocational College Construction Plan", which successively led to the establishment of 100 national model vocational colleges and 100 national key vocational colleges. In the past decade,

Table 5.2 List of 26 independent tourism vocational colleges (by June 2020)

No.	School name	Province	Location	Notes
1	Hebei Tourism Vocational College	Hebei	Chengde city	
2	Taiyuan Tourism College	Shanxi	Taiyuan city	
3	Tourism College of Shanxi	Shanxi	Taiyuan city	
4	Heilongjiang Vocational College of Tourism	Heilongjiang	Harbin city	
5	Shanghai Institute of Tourism	Shanghai	Shanghai municipality	
6	Nanjing Institute of Tourism and Hospitality	Jiangsu	Nanjing city	
7	Yangzhou Hospitality Institute	Jiangsu	Yangzhou city	Private
8	Jiangsu College of Tourism	Jiangsu	Yangzhou city	
9	Tourism College of Zhejiang	Zhejiang	Hangzhou city	
10	Zhoushan Tourism & Health College	Zhejiang	Zhoushan city	
11	Tourism College of Anhui	Anhui	Fuyang city	Private
12	Jiangxi Tourism and Commerce Vocational College	Jiangxi	Nanchang city	
13	Qingdao Vocational and Technical College of Hotel Management	Shandong	Qingdao city	
14	Shandong College of Tourism and Hospitality	Shandong	Jinan city	
15	Zhengzhou Tourism College	Henan	Zhengzhou city	
16	Zhengzhou Vocational College of Commerce and Tourism	Henan	Zhengzhou city	Private
17	Three Gorges Tourism Polytechnic College	Hubei	Yichang city	
18	Changsha Commerce & Tourism College	Hunan	Changsha city	
19	Hunan Golf and Tourism College	Hunan	Changde city	Private
20	Guangdong Vocational College of Hotel Management	Guangdong	Dongguan city	Private
21	Sanya Aviation and Tourism College	Hainan	Sanya city	Private
22	Hospitality Institute of Sanya	Hainan	Sanya city	Private
23	Chongqing Vocational Institute of Tourism	Chongqing	Chongqing municipality	
24	Tianfu New Area Aviation & Tourism College	Sichuan	Meishan city	Private
25	Yunnan College of Tourism Vocation	Yunnan	Kunming city	
26	Shaanxi Tourism Cuisine Professional College	Shaanxi	Xi'an city	Private

colleges supported by the project have made remarkable achievements in innovating system and mechanism, deepening teaching reform and improving service capability and teaching quality, which greatly optimized the environment of higher vocational education.

In 2010, the Ministry of Education, the Ministry of Human Resources and Social Security and the Ministry of Finance launched the plan for the construction

Table 5.3 List of tourism colleges and universities selected into the Plan for the Construction of National Model School of Secondary Vocational Education Reform and Development

School name (first batch)	School name (second batch)	School name (third batch)
Guangzhou Vocational School of Tourism and Business	Huhhot Vocational School of Tourism and Commerce	Changzhou Technical Institute of Tourism and Commerce
Guilin Tourism Vocational Secondary Professional School	Wuxi Higher Vocational School of Tourism and Commerce	Suzhou Tourism and Finance Institute
Haikou Vocational Tourism School	Cultural Tourism School of Pingdingshan City	Huangshan Vocational School of Tourism Management
Chongqing Tourism School	Huizhou College of Business	Zhangjiajie Vocational Tourism School
Xiamen Industrial-Commercial and Tourism School	Guizhou Tourism School	Hainan Provincial Tourism School
	Guangdong Provincial Vocational Tourism School	Panzhihua City Economic and Trade and Tourism School

of model schools in the reform and development of national secondary vocational education. It is planned to start the construction of about 1,000 model schools in the secondary vocational education reform and development in three batches of three years through standardization, informatization and modernization of these schools. These schools were to be built as exemplars for the reform and innovation, quality feature in national secondary vocational education and play a major leading role in the reform and development of secondary vocational education. This plan has greatly promoted the development of secondary vocational tourism education. Table 5.3 lists the tourism vocational colleges in this plan.

In 2015, the Ministry of Education issued the *Action Plan for Innovative Development in Higher Vocational Education (2015–2018)*, which proposed that the higher vocational education should take moral development as the foundation to serve the industry's development and promote employment. It should adhere to the principle of integrating education with industry development, learning with work, knowledge with practice and universities with enterprises. The action plan aims to build 200 quality vocational colleges and universities and enhance the quality of higher vocational education system in three years.

In March 2019, the Ministry of Education and the Ministry of Finance jointly issued the *Opinions on the Implementation of the Construction Plan of High-level Vocational Schools and Quality Programs with Chinese Characteristics* (referred to as the "Double High Plan"), and in December of the same year, the Ministry of Education and the Ministry of Finance released the list of construction plans of high-level vocational schools and quality programs with Chinese characteristics, which is an important system design in the field of vocational education after the "Double First-class" initiative in general higher education in China. A total of

197 vocational schools were selected in the first batch of "Double High Plan", of which 56 were in the construction of high-level schools and 141 were in the construction of quality program clusters. The "Double High Plan" is aimed to support the priority development of higher vocational colleges and programs with sound basic conditions, guide the vocational education to serve national strategies and finally build a batch of "model vocational colleges" that play an indispensable role in the local economy, are recognized in the industry and can be promoted through international communication.

At this stage, influenced by a series of quality projects such as the national demonstration construction projects (Table 5.4), tourism vocational education, both in vocational colleges and secondary vocational schools, emphasized on the construction of the key programs focusing on work-integrated learning.

In October 2009, with the support of China National Tourism Administration, China Union of Five-star Tourism Institutes (CUFTI) (CTI5) was established with five tourism colleges, namely Nanjing Institute of Tourism and Hospitality, Shandong College of Tourism and Hospitality, Tourism College of Zhejiang, Shanghai Institute of Tourism and Guilin Institute of Tourism. In 2013, Zhengzhou Tourism College also officially joined the alliance.

During this period, many colleges and institutes, such as Shanghai Institute of Tourism, Tourism College of Zhejiang, Nanjing Institute of Tourism and Hospitality, Shandong College of Tourism and Hospitality, Guilin Institute of Tourism and Yunnan College of Tourism Vocation, undertook some program construction projects both at provincial and ministerial levels, such as key construction programs, excellent courses, key construction teaching materials and model internship bases, which effectively promote the further development of the connotative construction of the higher vocational tourism education.

Table 5.4 List of tourism vocational colleges selected into the Projects for the Construction of Education Quality by the Ministry of Education

Project name	*Construction plan for national model (key) high vocational college*	*National quality vocational colleges (2019)*	*Construction of high-level vocational schools and quality programs with Chinese characteristics ("Double High Plan") (2019)*
Schools	1 Shanghai Institute of Tourism (2011) 2 Tourism College of Zhejiang (2013)	1 Tourism College of Zhejiang 2 Changsha Commerce & Tourism College	1 Tourism College of Zhejiang 2 Changsha Commerce & Tourism College 3 Qingdao Vocational and Technical College of Hotel Management 4 Hainan College of Economics and Business 5 Shaanxi Vocational & Technical College

In 2011, Shanghai Institute of Tourism became a national model vocational tourism college; in 2013, Tourism College of Zhejiang became a national model key vocational college; in 2013, the former Sichuan Cuisine College and the former Sichuan Agricultural Management College were merged to form Sichuan Tourism University; and in 2015, Guilin Institute of Tourism was upgraded to Guilin Tourism University.

With the development of both quality and coverage, tourism higher vocational education system has gradually become an important field for national tourism research. To some extent, the rapid and sustainable development of China's tourism industry benefits from the in-depth analysis of tourism phenomenon and the in-depth study of tourism theory in tourism colleges and universities.

Personnel training and social service

Provision of professional personnel for tourism

Tourism vocational education has made significant contribution in preparing tourism professionals needed in the industry and effectively solving the problem of insufficient professionals for the rapid development of tourism.

Tourism vocational education in China has roughly gone through three stages: beginning stage, building stage and developing stage. According to the statistics of the Personnel Department of the Ministry of Culture and Tourism, in 2017, there were 1,086 vocational colleges with tourism management majors and 947 secondary vocational schools with tourism majors in China, with more than 570,000 students. In the five years from 2013 to 2017, these colleges and schools produced about 1,084,900 graduates (Ministry of Culture and Tourism, 2018). Compared to the 215 tourism colleges and 49,022 students nationwide in 1990, tourism education has undergone tremendous changes in the past four decades. The system has provided a large number of professionals at all levels for the tourism industry, made great contributions to the expansion of China's tourism workforce, improved the quality of tourism practitioners and developed China's tourism industry. Tourism vocational colleges have become the main base in preparing professionals for China's tourism industry.

Contribution to social and economic development

Tourism has been playing an increasingly important role in the society and economy. Increasing leisure time among Chinese people, their growing desire in travel and increasing disposable income have made tourism an essential part of life. With the promulgation of relevant laws and policies, such as the *Tourism Law of the People's Republic of China*, the *Outline for National Tourism and Leisure* and the *Guidelines of the State Council on the Reform and Development of Tourism*, people's rights in travel and tourism have been fully acknowledged and legally protected. In this context, the significance of tourism vocational education to the development of the society and economy became more pronounced.

In the process of the Reform and Opening-up, shortage of talent in the workforce used to be a challenge. Tourism vocational education expanded learning opportunities for many people and effectively alleviated the common difficulties in employment in the society.

With the development of tourism vocational education, a group of professional tourism educators have emerged. Great achievements have been made in the teaching reform, scientific research and social services of tourism vocational education. The development of economy and society promoted the development of tourism, while at the same time, the development of tourism boosted the tourism vocational education, which, in turn, enhanced the transformation and upgrading of tourism and benefited the economic and social development.

Ways to serve the society

At present, the common forms of tourism vocational education to serve the society are as follows:

- Personnel training. It is the core of higher vocational colleges to cultivate application-oriented and highly skilled professionals for regions and industries, for example, to train professionals for new business type in the industry according to the changes of business, such as tourism e-commerce professionals.
- Training services. Vocational tourism education provides pre-job training, general post training and special training, such as training for scenic spot service personnel.
- Consulting services. Based on the characteristics of regional and industrial background, tourism vocational schools and colleagues provide enterprises with various application-oriented consulting services in decision-making, management and application technology, such as the formulation of tourism standards.
- Information services. With rich resources such as database and documents, tourism vocational schools provide information services for local industries and enterprises, such as the creation of tour itineraries and interpretation commentaries.
- Transformation of scientific and technological achievements, for example, the application of internet technology in the visitor flow control of scenic spots.
- Services to rural tourism development. The training for tourism practitioners is urgently needed due to the special form, resources and current momentum in the development of tourism in rural areas, such as in the transformation of rural scenic spots and the creation of B&B products. Tourism will play an important role in rural development and targeted poverty alleviation.

Distinct characteristics of tourism vocational education

The support and guidance from the national education and tourism administrative departments

The development of tourism vocational education in the past four decades has benefited from the rapid development of education and tourism in China, and it is also attributed to the support and guidance from the national education and tourism administrative departments.

In 2006, the Ministry of Education established the "Teaching Supervisory Committee for Tourism Discipline in Higher Vocational College" and "Teaching Supervisory Committee for Cuisine Program in Higher Vocational College". In the same year, the Ministry of Education also launched the "Plan for the Construction of National Model Higher Vocational College" and built 100 national model higher vocational colleges, among which Shanghai Institute of Tourism was listed. In 2010, the "Plan for the Construction of National Key Higher Vocational College" was launched, with about 100 vocational colleges added into the list on the basis of the original National Model Higher Vocational Colleges. Tourism College of Zhejiang was among the added vacation colleges.

In 2002, China National Tourism Administration promulgated the *Outline of the Training of Professionals for China's Tourism industry in the "Tenth Five-year" Plan*. In 2008, the Educational Branch of China Tourism Association was established under the guidance of China National Tourism Administration. In 2006, China National Tourism Administration established its research grants scheme. In 2013, the Training Plan for Young Experts for Tourism Industry was started, followed by the Ten Thousand Tourism Experts Plan in 2015. In such a context, tourism vocational education has become one of the main platforms for training skilled tourism personnel in the country.

It is worth mentioning that in 2015, China National Tourism Administration and the Ministry of Education jointly issued the *Guidelines of Accelerating the Development of Modern Tourism Vocational Education*, which is the first policy document on the development of tourism vocational education jointly issued by the two ministerial agencies. Thanks to the relevant national policies, tourism vocational education gained rapid development and quickly embarked on a standardized, orderly and healthy path for development.

The early tourism colleges and universities were run by China National Tourism Administration or provincial tourism administrations, and most of the industry training for tourism practitioners was also carried out by tourism authorities or tourism enterprises. Tourism education has a close relationship with industry enterprises from the very beginning. For example, the five colleges or institutes of the CUFTI were all founded by tourism administration departments.

The cooperation between industry and colleges is an embodiment of "developing majors around tourism, cultivating professionals for tourism, conducting scientific research about tourism and providing services to tourism" in the

Guidelines. Most tourism programs are set up exactly to the needs of tourism development, so do the scientific research and social services.

The cooperation between industry and colleges has created a team of "dual-profession" teachers. Tourism education originated from the development of tourism. In the early stage, some teachers in tourism vocational education came from the fields of hotel, travel agency or catering industry. This tradition remained until the present day, making "dual-profession" teachers a unique feature in tourism vocational education.

Economic globalization accelerates the internationalization of tourism and promotes the internationalization of tourism education. At the same time, internationalization of tourism presents challenges for the requirements and training methods of tourism education. Many colleges have actively explored effective internationalization models, mainly through Chinese-foreign cooperation, introducing international advanced education mode and participating in international education quality certification. For instance, China-Australia International Hotel Management College was jointly operated by Tourism College of Zhejiang and William Angliss Institute of Australia; Tourism College of Zhejiang and Nanjing Institute of Tourism and Hospitality have been accredited by the TedQual certification system of UN World Tourism Organization (UNWTO).

Challenges of tourism vocational education

The appealing power of tourism industry employment has declined. Jobs in the tourism industry appear less attractive in the job market. One reason is the low payment level in tourism jobs. According to the *Average Wage of Different Posts in Online Direct Reporting Platform Enterprises* released by the National Bureau of Statistics in 2016, the average annual income of the employees in hotel and catering industry is RMB 40,573, ranking last among all industries, accounting for only 70.7% of the all-industry average income of RMB 57,394 (National Bureau of Statistics, 2017).

The competition for students among vocational colleges has intensified. General high school graduates are always the main target of recruitment of higher vocational colleges. According to the Ministry of Education, the number of applicants for the college entrance examination reached a historical peak of 10.5 million in 2008, followed by the constant decline the following five years to only 9.4 million in 2016. In terms of admission rate, it has been on the rise every year starting from 57% in 2008, with the highest admission rate reaching 81.13% in 2018 (Ministry of Education, 2017).

Secondary vocational school graduates present another basic source of recruitment for higher vocational colleges. From 2009 to 2012, the number of secondary vocational school graduates was on the rise. The number of secondary vocational school enrollment has been decreasing after reaching the highest point in 2011. Generally speaking, secondary vocational schools also saw a decrease in student numbers.

The number of college enrollment has enjoyed a steady growth since 2009, despite the declining base for student recruitment. Taking higher tourism vocational colleges as an example, in 2016, there were 1,086 vocational colleges offering tourism programs, accounting for 80.0% of 1,359 vocational colleges in China. The top three programs in terms of enrollment are Tourism Management (set up in 826 colleges, accounting for 60.8%), Hotel Management (in 668 colleges, accounting for 50%) and Tourism English (in 181 colleges, accounting for 13.3%) (Ministry of Education, 2017). Tourism Management has become one of the most popular programs in China. The increase of colleges and universities with the above-mentioned programs and the relatively stable number of applicants resulted in a significant reduction in the number of students allocated to colleges and universities in various provinces and cities. In order to attract students, competitions among colleges are becoming even fiercer, consequently making it even harder to cultivate quality graduates.

Generally speaking, the number of potential students of higher tourism vocational colleges tends to decrease and so do the number of students in secondary tourism vocational colleges. This leads to the shortage of students in higher vocational colleges coming from secondary vocational schools. The decrease of potential students presents a serious practical challenge for tourism vocational education in China.

The level of recognition of vocational tourism education by the society, parents and students is low. In China, it is commonly believed that the longer one's education period goes, the better the outcome, causing the underestimation of the role of vocational education. In particular, vocational education mainly operates at the junior college level and is thus regarded as low on the education ladder.

The modern vocational education system is yet to be improved. In order to establish an effective modern vocational education system, the relevant departments of the State successively promulgated the *Outline of China's National Plan for Medium and Long-term Education Reform and Development (2010–2020)* in 2010, the *Decision of the State Council on Accelerating the Development of Modern Vocational Education* in 2014 and the *Construction Plan of Modern Vocational Education System (2014–2020)* (hereinafter referred to as the *Plan*). These policies draw the blueprint of an integrated modern vocational education system. In terms of the types of vocational education, the *Plan* proposed that "both diploma and non-diploma vocational training should be undertaken by vocational colleges to meet the diversified needs of industries, enterprises and communities". With regard to the grades structure of vocational education, the *Plan* proposed that the vocational education system would include primary vocational education, secondary vocational education and higher vocational education. Among them, higher vocational education should focus on developing colleges and universities of applied technology to cultivate professionals at the undergraduate level, on the basis of the existing higher vocational (schools) colleges.

However, there is a big gap between the reality of tourism vocational education and the above requirements. First, the scale and number of colleges of non-diploma vocational education are small and not comparable to diploma level vocational education, and thus cannot meet the needs of tourism development for professionals. Second, the current tourism vocational education system mainly includes secondary vocational education and higher vocational education, but lacks undergraduate degree level or above programs. Although there are a few undergraduate tourism programs in hotel management or tourism management, they are not connected with secondary vocational and higher vocational tourism programs. Third, the future vocational education system needs to be improved, considering that the current tourism vocational education system is not meeting the needs of tourism development for professionals. Of course, it is not a task that can be accomplished only by the reform of tourism vocational education. It also needs the whole vocational education system to be reformed along the planned route.

Opportunities of tourism vocational education

As the role of tourism in economic and social development becomes increasingly important, tourism vocational education has great potential for its further development. The future tourism vocational education should pay more attention to the quality of education through deepening the integration of industry and education.

A shift from "quantity" to "quality". China's economic development has entered a new era, basically featuring a shift from high-speed growth to high-quality development. At a time of promoting the country from a big tourism country to a strong tourism country, quality and efficiency should be taken as the priority. In particular, the proposal of "China Service" and its implementation would present a huge demand on quality tourism professionals. China has formed the largest tourism education system in the world, which should focus more on "quality" than "quantity" in its future development. In the next few decades, several "Chinese Brand" tourism colleges and programs should make their debut on the world stage of tourism education.

A change from "general cultivation" to "characteristic training". The characteristics of tourism industry determine that tourism vocational education should have flexible training mode, so as to be adaptable to local conditions and serve the industry. In the past, tourism education emphasized on standardization and the unified mode of training, indicated by a series of professional standards, practice standards and requirements on teaching material issued by the Ministry of Education and the National Tourism Administration. Today, the tourism industry is diversified with constantly emerging new business forms. Flexible education system should be implemented in the cultivation of professional personnel to meet the changes in the market and the needs of the society. Various training modes should be adopted to cultivate the practical ability of the students through practice-based teaching and learning.

Tourism education and tourism industry going from "cooperation" to "integration". It is urgently required that structural reform of the human resource supply be promoted to deepen the integration of industry and education. Educational Institutions and industrial enterprises should gradually progress from the original "cooperation" to "integration" to get the industry more involved in the education process.

Scientific research goes from "interpretation" to "innovation". Tourism scientific research grows up with the development of tourism and tourism education. In the past four decades, the majority of tourism scientific research in China has taken "interpretation" as the main approach. On one hand, western tourism research models and methods have been used to "interpret" China's tourism phenomena and industrial development; on the other hand, the models and methods for theoretical research in other disciplines have been used to "interpret" tourism phenomena and tourism behaviors. Today, we should go from "interpretation" to "innovation" in the scientific research for tourism, for "innovation" has become the important task of scientific research. More attention should be paid on the research of basic tourism theory. At the same time, attention should be paid to using tourism theory to guide innovation practices. Tourism scientific research should not be purely theoretical research, but be applied as research closely related to industry problems. Whether it can guide the development of tourism industry, solve the industry problems and explain the nature of tourism activities is an important criterion to test the effects of research.

Conclusions

This chapter mainly analyzes the system, role, contribution and characteristics of China's tourism vocational education. The 40 years of development of China's tourism vocational education witnessed great achievements in operational models, school-enterprise cooperation, teacher training, curriculum development, teaching reform and so on. The development of China's tourism vocational education was closely related to the economic development and social progress of the whole country. However, there is still a long way ahead to develop a modern tourism vocational education system. For example, China's tourism education is essentially based on the hierarchical education system of doctoral, master-level, undergraduate, higher vocational and secondary vocational education, which consequently leads to the fact that tourism vocational education cannot enjoy the equal status with general education. However, it is believed that with China's new orientation towards the global market, the determination of the country to speed up the development of modern vocational education and the promotion of the "China service" ideology, the development of China's tourism vocational education will also continue to be a significant part in serving the country's admirable tourism industry development.

References

Ministry of Culture and Tourism (2018). *Statistics on national tourism education and training 2016*. Available from: http://zwgk.mct.gov.cn/auto255/201804/t20180419_832575.html?keywords= [22 March 2020].

Ministry of Education (2016). *The plan for poverty alleviation through education during the "13th Five-Year" plan*. Available from: http://www.moe.gov.cn/srcsite/A03/moe_1892/moe_630/201612/t20161229_293351.html [20 February 2020].

Ministry of Education (2017). *Official report on statistics of national education development 2016*. Available from: http://www.moe.gov.cn/jyb_sjzl/sjzl_fztjgb/201707/t20170710_309042.html [18 February 2020].

Ministry of Education (2021). *The Catalogue of Programs of Vocational Education (2021)*. Available from: http://www.moe.gov.cn/srcsite/A07/moe_953/202103/t20210319_521135.html [20 March 2020].

National Bureau of Statistics (2017). *Average wages of different positions of online direct reporting platform enterprises in 2016*. Available from: http://www.stats.gov.cn/tjsj/zxfb./201705/t20170527_1498364.html [20 February 2020].

UNESCO (2011). *International standard classification of education ISCED 2011*, United Nations. Available from: http://uis.unesco.org/en/topic/international-standard-classification-education-isced [23 February 2020].

Wang, K. (2018). The training system of tourism professionals with Chinese characteristics has been formed. *China Tourism News*, 10 October, p. 4. Available from: http://news.ctnews.com.cn/zglyb/images/2018-10/10/04/ZGLYB2018101004.pdf [16 February 2020].

6 Tour guiding education, training and administration

Xianrong Luo, Hanjun Tao and Lili Liu

Introduction

This chapter mainly introduces the development of tour guiding education in Mainland China, including how the universities were involved in the process since China's reform and opening up, and the content and implementation strategies for tour guiding training and administration. The preparation of tour guides is important for the conduct of many tourism activities and thus for the development of tourism industry as a whole (Artemyev, Abdreyeva & Batbaatar, 2018). The extant literature on tour guiding mainly saw tour guides as service providers, cultural interaction facilitators, tourism mentors and so on (Chen & Chang, 2019). While studies investigating the behaviors and managerial issues of tour guides (Huang, 2010; Song & Wang, 2013) can be found, little research has concerned how to well prepare the tour guides from the perspectives of education and training.

This chapter includes two sections: "tour guide education and training" and "tour guide administration". The chapter aims to provide a general understanding of the tour guide education and administration in China. Data from the China National Tourism Administration (CNTA, now merged into the Ministry of Culture and Tourism) and from the Internet was mainly used. One former officer of CNTA, who was in charge of tour guiding education, and the general secretary of China Tourism Education Association also provided considerable information for the writing of this chapter. Tourism in recent years played a significant role in improving people's quality of life, particularly with the judgment of the 19th National Congress of the Communist Party of China that the society's main contradiction of China has transitioned into the contradiction between "people's increasing demand on high-quality life and unbalanced and insufficient development" (Zhao, 2018). Also, China has become a major actor to influence the development of international tourism, especially as a tourist-producing country. Therefore, this chapter, by introducing the system of tour guides education, training and administration in China, helps to understand the recent tourism development in China.

Tour guide education and training

There was no tour guide education course in the universities of China before 1978, although the Chinese International Tour Service as a travel agent was established

DOI: 10.4324/9781003004363-6

in 1954. At the time, some graduates from foreign language schools were selected to receive foreign guests and to provide interpretation service. These graduates, nevertheless, received little or no tour guiding training. After the opening policy of China in 1978, international tourists visiting China increased significantly, which helped to promote the construction of tourism facilities, and the establishment of tourism schools and the development of tour guiding courses.

Tour guiding education in schools and universities/colleges

Tour guiding education in China began in 1979 when the Shanghai Institution of Tourism set up "English tour guiding and translation" as a teaching subject. In the same year, Dalian University of Foreign Languages also started to enroll students studying "Japanese tour guiding and translation". Later in the 1980s, Xi'an University of Foreign Languages, Beijing Institute of Tourism (now the Institute of Tourism of Beijing Union University) and Changchun University also started their tour guiding education in foreign languages.

Tour guiding education in China at the tertiary level includes that in universities and that in higher vocational colleges. In 2015, there were 102 universities and colleges which had tour guiding education as a study field, with a total enrolment of 3,889 students. Apart from that, there were 116 universities or colleges which set up disciplines closely related to tour guiding education. In 2015, these universities or colleges enrolled 3,596 students. Tour guiding education at the secondary level mainly refers to that in technical secondary schools and vocational high schools. In 2015, these schools altogether enrolled 11,145 students to study in tour guiding related programs ("Tourism related foreign language" and "Tour guiding service") and 12,361 students graduated from these programs. These students normally have to study two to three years.

Tour guiding education at the tertiary level mainly includes courses such as tour guiding basic knowledge, basic facts of tourist destination or generating countries, tour guiding practices and skills, tourism related policy and regulations and simulated tour guiding. In the course of tour guiding basic knowledge, students were expected to build up a general understanding of the Chinese history and culture, referring to the main landscapes, architectures, gardens, religions, food cultures and literacy works in China. Countries, mainly those in Asia, Europe, North America and Oceania, were introduced as tourist-generating countries to the students. The students were also required to grasp the general tour guiding practices and skills, such as the general procedure of service provision to tourist groups or individual tourists, tour guiding language and interpretation skills and related insurance and sanitation knowledge. In order to help students better practice tour guiding in reality, a range of laws and regulations were also introduced and taught in the guiding programs, and simulated tour guiding was normally organized as part of the tour guiding education in universities and colleges.

Tour guiding education at secondary level also includes courses such as tour guiding basic knowledge, tourism-related policy and regulations. Yet, such courses tend to be simple and less critical in contents, usually including memorable facts and requiring little reasoning skills. To teach the students tour guiding skills, a range of courses like the art of tour guiding for tourist groups, road commentaries and interpretation at scenic spots are organized. Some schools also set up courses to shape the body posture for the students, to improve their etiquette and speaking skills and to regulate their ethical behavior while providing tour guiding services.

Though different, tour guiding education at tertiary level and at secondary level both aim to prepare students with the capabilities of providing practical tour guiding services. Students are taught with theoretical knowledge as well as practical skills. Apart from finishing the courses in school or university/college, students also have to work as an intern in the industry for at least three months. In some universities or colleges, students are required to attend the national tour guiding qualification examination and to get the certificate for taking tour guiding jobs in the future.

Tour guide training

Tour guides can be divided into different types. Based on the services they provide, tour guides are categorized into outbound tour leaders, national guides, local guides and interpreters at scenic spots. Some tour guides are only qualified to provide tour guiding service in Chinese, while others could do their job in Chinese as well as in foreign languages. Based on the level of skills and professional experience, tour guides in China are classified into four professional levels: junior, medium, senior and specialist. Based on the type of employment, they could be divided as full- or part-time tour guides for travel agents or self-employed tour guides.

The CNTA paid great attention to the work of tour guide training. CNTA issued a number of notices to push local tourism departments and tourism agents to train and manage the tour guides under their jurisdiction and particularly to improve their capacity in providing quality tour guiding services. For example, in 2000 CNTA issued the *Notice to Enhance and Improve the Work of Tour Guide Training*, according to which local tourism departments have to make an annual plan for tour guide training and to well monitor the local work of tour guide training; travel agents also have to regularly train their tour guides, particularly in the low season, and to record the training experience for each tour guide; travel agents should also reward employees who have regularly attended tour guide training. To further highlight the importance of tour guide training, the notice clearly requires that every tour guide attend training for at least 56 hours per year.

There are two types of tour guide training: pre-vocational training and in-service training. The former is for those who have not gained the qualification for providing tour guiding services while the latter is for those who have already gained the qualification.

Pre-vocational training

Pre-vocational tour guide training includes the training needed for passing the national tour guiding qualification examination and for providing real tour guiding services. The national tour guiding qualification examination is held once a year. Chinese citizens, who are healthy, have finished their high school education or secondary vocational education or higher levels of education and have the basic knowledge and language skills to be a tour guide, can attend the examination and may gain the qualification of providing tour guiding service. The examination has two parts, separately known as written examination and oral test. Content for the examination generally refers to four aspects: tour guide basic knowledge, tourism-related policy and regulations, tour guiding practices and simulated tour guiding, which are very similar to the content of tour guiding education in schools and universities/colleges. Now, in China, there are a range of training agencies which are allowed to provide training courses for people who plan to attend the national tour guiding qualification examination. Local tourism departments also arrange some training courses for the people who are attending the examination.

After gaining the qualification of providing tour guiding services, an individual also has to attend the training courses arranged by local tourism departments (for at least a week) before becoming a practicing tour guide. Content of the courses normally consists of professional ethics and career plan for tour guide, general preparations to be a tour guide and the procedure of tour guiding and related issues. Only after passing these courses can an individual work as a practicing tour guide.

In-service training

In-service training aims to improve the quality of service that tour guides provide and is mainly carried out by travel agents and local tourism departments. In-service training can be categorized into general training and special training. General training aims to reinforce tour guides' professional ethics, knowledge of policy and regulations and business skills. For example, if a junior tour guide wants to be promoted to a medium-level guide, he/she has to attend such general training and then pass related exams. Special training, however, is designed for some tour guides to learn particular skills or knowledge. It is normally organized in the following situations:

- Travel agents organize special training, particularly for tour guides who are short of certain knowledge or skills; individual tour guides who feel insufficient in certain knowledge or skills could also attend related special training organized by some institutions.
- Local tourism departments decide to train tour guides who work under their jurisdiction in order to advance their knowledge or skills for solving

particular problems. Content for the training is generally designed based on surveys. For example, Beijing Municipal Tourism Development Commission organized training for those tour guides who could not provide proper interpretation for tourists visiting the Imperial Palace. In the implementation, some experts and advanced tour guides were invited to give lectures; the training lasted continuously for six periods and in each period about 100 tour guides attended; training contents include the history and architecture of the Imperial Palace, traditional handicrafts related to the Imperial Palace, the development of Jade articles in Ming Dynasty and so on.

- The national tourism department, previously CNTA and now the Ministry of Culture and Tourism, requires provincial and local tourism departments organize special training in certain situations. For example, to deal with the issues arising from the increasing media exposure of uncivilized behaviors of Chinese outbound tourists in recent years, local tourism departments were required to organize training which aimed to help outbound tour leaders to properly guide the behavior of outbound tourists. Such training contents include "regulations for tour leaders to guide tourists' behavior", "regulations to educate tourists before traveling" and "instructions to practically regulate tourists' behavior". The training aimed to first regulate the behavior of tour leaders themselves and then to increase their ability in managing tourist groups.

The training introduced above, either pre-vocational or in-service, is important in keeping the capacity of the tour guides workforce. Pre-vocational training is normally organized before the national tour guiding qualification examination that is held annually in November and December each year. In order not to interfere in tour guides' work, in-service training is often organized during the off seasons, generally in winter or early spring each year. Either pre-vocational training or in-service training can be organized in different ways by tourism departments or by institutions or by training centers. In recent years, with the development of Internet technology, online training for tour guides is becoming increasingly popular.

Tour guide competitions at national level have been organized regularly in order to advance the capacity of the tour guide workforce. For example, CNTA, together with the Central Committee of the Communist Youth League of China and the All-China Women's Federation, organized the second national tour guide competition in 2012; all qualified tour guides who had worked for one year or more could attend the competition. The competition had three rounds. While the first and second rounds were organized by provincial tourism departments, the third round was organized by CNTA. Tour guides who obtained good results in the competition could be promoted to higher professional levels. To prepare for a competition, local tourism departments organize different forms of training for tour guides, which strengthen the capacity of the tour guides workforce. Additionally, China Tourism Association also regularly organizes competition for tourism students majoring in tourism in order to prepare them for their work after graduation.

Tour guide administration

The issue of tour guide administration is key to the development of the tourism industry and specially to keep the image of destinations and travel agents. To manage the workforce of tour guides, service standards, registration and management system and related policies and laws were particularly formulated in the administration system.

Service standards for tour guides

In order to support and supervise the work of tour guides, service standards for tour guides in various situations are designed and promoted. The main service standards for tour guides to follow in varying settings are listed in Table 6.1. The standards, on the one hand, instruct tour guides and travel agents about

Table 6.1 Main standards for tour guide service

Number	Name	Standard no.	Main content
1	Tour guide service standards	GB/T 15971-2010	Regulations for the basic qualities a tour guide needs, the procedure of providing tour guiding service, what kinds of services and how to solve the general problems during the process.
2	Domestic service standards for travel agents	LB/T 004-2013	Regulations for the quality of domestic tourism service that travel agents provide, including the quality of tour guiding service.
3	Outbound service standards for travel agents	LB/T 005-2011	Regulations for necessary products and the quality of outbound tourism service that travel agents provide, including the quality of tour guiding service that tour leaders should provide and their responsibilities.
4	General service standards for travel agents	LB/T 008-2011	General regulations for travel agents to provide tourism service, including how to manage tour guides and tour leaders.
5	Inbound service standards for travel agents	LB/T 009-2011	Regulations for inbound tourism services in terms of product development and marketing, procedure of service provision and service practices. Regulations particularly refer to the responsibilities of tour guides and the quality of tour guiding service.
6	Service standards for interpretations at scenic spots	LB/T 014-2011	Regulations for the quality of interpretation service at scenic spots. Requirements for those interpreters are consistent with that for tour guides.

(Continued)

Number	Name	Standard no.	Main content
7	Security regulations for travel agents	LB/T 028-2013	General regulations for the security issue of travel agents, including how to guarantee security in daily management and business operation. To make the practices of tour guides secure is also a key issue.
8	Regulations for student internship at travel agents	LB/T 032-2014	Setting the general principles for the university/school, the student and the travel agent to direct the internship process; requirements for the work of the (intern) position are particularly formulated.
9	Regulations for guiding tourists to behave politely	LB/T 039-2015	Setting the basic requirements, detail content and regulations for tour guides or tour leaders in how to properly guide the behavior of tourists.
10	Service standards for travel agents to serve senior tourists	LB/T 052-2016	Regulations for providing services to senior tourists, including requirements for tourism products and tourist marketing, implementation of travel plans and the service practices of tour guides or tour leaders.
11	Regulations for receiving student tourists from Hong Kong and Macao	LB/T 053-2016	Setting the procedure and general regulations for providing services to student tourists from Hong Kong and Macau. The quality of tour guiding service is particularly mentioned.
12	Regulations for providing study trip service	LB/T 054-2016	Setting the basic requirements for providing study trip services, including how to design the products and how to provide the services. The quality of interpretation by tour guides or interpreters is particularly mentioned.

how to provide quality services and, on the other hand, make their work easier to be evaluated. Additionally, when new problems arise (e.g., impolite behaviors of tourists and the security issue of senior tourists) or certain tourism products become popular (e.g., the popularization of study tour trips), related regulations may be revised to direct the work of tour guides/tour leaders.

Tour guide management system

Since the 1980s, the scale of the tour guide workforce in China has become larger and larger, along with the development of the tourism industry. To manage the huge number of tour guides, CNTA developed a management system, which mainly referred to two aspects: (1) how to become a tour guide; and (2) how to upgrade to be an advanced tour guide.

How to become a tour guide

As mentioned above, those people who want to become a tour guide have to attend the national tour guide qualification examination. Passing the qualification examination is the first indispensable step if a person wants to be a tour guide. The examination could be conducted in Chinese or other foreign languages such as English, Korean, Japanese and so on. Accordingly, people who have passed the examination can gain the qualification to work as a Chinese tour guide or a tour guide who work in certain foreign languages.

Once a person has passed the examination and gained the qualification of being a tour guide, he or she has to attend certain pre-service training and then register to be a practicing tour guide, either through a travel agent or a tour guide service company. People who provide tour guide services without obtaining the qualification and fulfilling the processes are regarded as illegal tour guides.

How to upgrade to be an advanced tour guide

There are four levels of tour guides: junior tour guide, medium tour guide, senior tour guide and specialist tour guide, respectively. CNTA promulgated the standards to evaluate these tour guides at different levels. Accordingly, tour guides at different levels differ in terms of knowledge, skills, education level as well as experience.

Junior tour guides: They are required to understand related policies and regulations, have basic knowledge about the main local scenic spots and provide general interpretations, have general knowledge about China's politics, history, geography, religion and customs, and generally understand the conditions and customs of tourist origin countries/regions. For Chinese junior tour guides, they have to gain proficiency in mandarin, at least at a high school graduate level; for junior tour guides using foreign languages, they have to grasp one foreign language and can use it at a third-year university student level specializing in the language. In terms of skills, they need to complete related tourist reception work individually, build harmonious relationships with tourists and properly report their work to their supervisors. They are supposed to conduct their work without incurring serious complaints from tourists. To be a junior tour guide, people have to pass the national tour guide qualification examination first and then work in the tourism industry for at least one year.

Medium-level tour guides: Compared with junior tour guides, they have to be more familiar with related policies and regulations, have comprehensive knowledge about the main local scenic spots and could provide an overall interpretation, be familiar with China's politics, history, geography, religion and customs, the conditions and customs of tourist origin countries/ regions, and have a good understanding of the working regulations for tour guides. For Chinese-language medium-level tour guides, they have to have language proficiency in mandarin at least at a college graduate level; for medium-level tour guides using foreign languages, they have to specialize in the foreign language and can use it at least as a university graduate who majors in the language. In terms of working skills, they need to be capable in serving various tourist groups and provide good tour guiding services; they should be able to solve the various problems occurred while providing services; they are required to work properly together with tourists, related companies and other employees. Chinese-language medium-level tour guides are also supposed to understand at least one dialect and medium-level tour guides using foreign languages are supposed to provide correct interpretations. Medium-level tour guides should also be able to mentor junior tour guides. To be a medium-level tour guide, the candidate has to work as a junior tour guide for at least three years.

Senior tour guides: Compared to junior and medium-level tour guides, senior tour guides need to be more familiar with and have a better understanding of related policies and regulations, to have comprehensive and in-depth knowledge about the main local scenic spots and a good understanding of the conditions and customs of tourist origin countries/regions. For Chinese-language senior tour guides, they have to be able to work in mandarin as well as one dialect; for senior tour guides using foreign languages, they need to work expertly with one foreign language (being capable to provide general oral interpretation) and also primarily use another foreign language. While providing tour guiding services, senior tour guides should solve related problems or even emergencies properly and provide quality and creative interpretations. They are expected to have a good reputation among peers and in the industry as well, and to possibly conduct some research work in relation to tour guiding. To be a senior tour guide, one has to work as a medium-level tour guide for at least three years.

Specialist tour guides: In terms of the above-mentioned policies and regulations, knowledge of main scenic spots and conditions and customs of tourist-generating areas, specialist tour guides are required to have an in-depth understanding. They should be able to solve the problems at work properly and provide accurate and insightful interpretations. For Chinese-language specialist tour guides, they have to be able to work in mandarin as well as in one dialect; for specialist tour guides using foreign languages, they need to work expertly with one foreign language (e.g., being able to translate expertly for tourism conferences and other related settings) and can use a second foreign language. They are expected to develop scripts of interpretation by themselves and able to train and mentor senior tour guides. The number of specialist tour guides in the country is small. They generally have very good reputation and are consulted on the development of China's tourism industry. Specialist tour guides normally have some published works related to tour guiding. To be eligible for applying for a specialist tour guide, one has to work as a senior tour guide for at least five years.

Practically, when a junior tour guide applies for medium tour guide qualification or when a medium tour guide applies for senior tour guide qualification, he or she has to attend related writing tests. Before attending the tests, he or she needs to provide tour guiding services at least for 90 tourist groups. When senior tour guides apply for specialist tour guides, they need to have very good professional performance and reputation; they also need to have some publications and pass a panel assessment for thesis defense.

Tour guide management: laws and regulations

Some laws or regulations have been formulated in managing tour guides. These include the *Tourism Law of the People's Republic of China, Travel Agency Regulations*, and *Regulations on the Administration of Tour Guides*. The *Tourism Law of the People's Republic of China* was promulgated in 2013. Apart from some general issues, the law requires tour guides and tour leaders to wear related qualification certificate ID card when working, provide tour guiding services ethically, respect the customs and religions of tourists, and guide tourists to behave properly. While guiding a tourist group, tour guides or tour leaders should neither change the schedule nor stop providing related services; they should not coerce tourists for shopping against their will. Demanding for tips is strictly forbidden. The *Travel Agency Regulations* was published in 2009 in order to regulate the business of travel agencies and to protect the rights of tourists. Under these regulations, tour guides are required to provide tour guiding services according to related contracts, follow the agreed tour itinerary strictly except in the cases caused by forces beyond human control. Tour guides are not allowed to engage in coerced shopping and ask for tips against tourists' will. According to the *Travel Agency Regulations*, tour guides and tour leaders are responsible to protect the tourists they provide services for and particularly need to report to police officers and the embassy (when travel overseas) when there are dangers; for outbound tourists who illegally overstay in a foreign country or inbound tourists who overstay in China, tour guides or tour leaders are responsible for reporting the cases to relevant government departments. The *Regulations on the Administration of Tour Guides* was published in 1999 to regulate the service-providing process of tour guides and to protect the right of tourists as well as that of tour guides. It regulates the rights, responsibilities, liabilities and related penalties when certain situations happen.

The above-mentioned law and regulations provide clarifications on tour guides' responsibilities or professional requirements. The rights and welfare of tour guides nevertheless remain to be further clarified. In 2017, the *Tour Guide Management Measures* was promulgated as a policy to protect the rights and welfare of tour guides specifically.

Conclusions

This chapter elaborated on the education, training and management system of tour guides in China. In recent years, the tourism industry in China has experienced rapid development. According to the national tourism statistics, in 2018 China recorded around 5.5 billion domestic tourist trips, around 14 million inbound tourist arrivals and 15 million outbound tourist departures (Ministry of Culture and Tourism of the People's Republic of China, 2019). To support such a scale of development of the tourism industry, tour guide education, training and management appears to be a significant issue.

In the literature, problems that tour guides encounter and conflicts they might have with travel agencies have been discussed (Wang, 2009; He, Xiao & Liu, 2010). Scholars also mentioned some unethical behaviors of tour guides which are seen as a result of insecure working conditions and low income (Chen, 2006). Therefore, providing education and training to tour guides in dealing with these issues and protecting the rights and welfare of tour guides are both important issues. As mentioned above, tourism authorities increasingly realize the importance of tour guides' rights and welfare and are trying to improve their working conditions through establishing a better regulation and management system.

This chapter, by introducing the education, training, and management system of tour guides in China, provides a foundation for understanding the work and life of tour guides in China. Working as a tour guide seems not easy, given the training and examination one has to undertake and the professional responsibilities in a tour guide's role. Nevertheless, incomes for tour guides seem to be unstable and largely dependent on how much commission they could obtain from tourists' on-site expenses. To better understand their work and life conditions, more studies should be conducted to investigate the income and welfare of tour guides in China.

References

Artemyev, A., Abdreyeva, S., & Batbaatar, Z. (2018). Tour guiding as a factor of tourism development along the route of the Great Silk Road in Kazakhstan. *Advances in Economics, Business and Management Research, 11*, 1–4.

Chen, T. (2006). On safeguarding the professional rights and interests of tour guides and their benefit expression (in Chinese). *Tourism Tribune, 21*(4), 60–66.

Chen, J., & Chang, T. C. (2019). Touring as labour: Mobilities and reconsideration of tour guiding in everyday life. *Tourism Geography, 22*(4–5), 813–831.

He, A., Xiao, Z., & Liu, S. (2010). An analysis of tour guides' moral risks based on dynamic optimization model (in Chinese). *Tourism Tribune, 25*(9), 65–70.

Huang, S. (2010). A revised importance-performance analysis of tour guide performance in China. *Tourism Analysis, 15*(2), 227–241.

Ministry of Culture and Tourism of the People's Republic of China. (2019). The 2018 statistics bulletin for China's culture and tourism development. Accessed on August 23 of 2020 at http://zwgk.mct.gov.cn/auto255/201905/t20190530_844003.html?keywords=.

Song, Z., & Wang, Y. (2013). Analysis on the access system and management of tour guides (in Chinese). *Tourism Tribune, 28*(7), 57–63.

Wang, C. (2009). On the expurgation of legal relations between part-time tour guides, tour guide service companies and travel agencies (in Chinese). *Tourism Tribune, 24*(11), 64–70.

Zhao, X. (2018). Practice and innovation in 'Science of Tourist Guide' curriculum reform: Summarizing classroom teaching reform experience and implementing the spirit of the 19th National Congress of the Communist Part of China from practice. *Advances in Social Science, Education and Humanities Research, 151*, 36–40.

7 Master of tourism administration education

Yanbo Yao and Pingping Hou

Introduction

In line with the development trend of tourism higher education worldwide and the urgent need for high-level tourism professionals with practical skills in China, the Ministry of Education of China designated for the first time in September 2010 that some universities in China have the qualification of establishing Master of Tourism Administration (MTA) education program, which marks a new stage in China's Master degree education in tourism studies. This chapter introduces the inaugural background, development routes, and enrolment size of the MTA program in China. The presentation of the chapter will revolve around the following four aspects of MTA education: operational modes, curriculum structure, assessment, and problems and issues. There are several distinct characteristics in China's MTA education program: diversified school-running models, gradual improvement of the curriculum system and professional settings, the teaching model of applying theory into practice, diversified student evaluation systems, innovative mode of teaching, and the interactive development of internationalization and localization. China's MTA education program can be improved from the aspects of innovating curriculum content, enhancing teacher's abilities, and compiling high-quality teaching materials.

Development routes of MTA education

Settings of MTA education

With the rapid development of economy and society, the adjustment of industrial structure as well as the transformation of economic mode, social demand for high-level tourism talents is growing rapidly. Professional degree education with characteristics of professionalism and application ability is gradually recognized by all sectors of society. However, the postgraduate education system oriented by the cultivation of academic skills in China is far from meeting the needs of social and economic development in industry sphere. Therefore, it has become an urgent need to adjust the industrial structure to strengthen the cultivation of talents with applied skills. In view of this, the Ministry of Education decided

DOI: 10.4324/9781003004363-7

to accelerate the development of master degree education since 2009 to further accelerate the pace of postgraduate education structure adjustment and enhance the quality of postgraduate education for better serving the country's economic and social development. In 2010, the professional master program was put on the position as important as the academic master, which marks a change of postgraduate education in China from focusing on the cultivation of academic talents to the cultivation of versatile and practical talents.

In 2009, the State Council established tourism as a national strategic pillar industry and promoted the upgrading of tourism sector from a traditional industry to a modern service industry, which requires large numbers of versatile and practical professionals to join the tourism industry. However, the discrepancy between the demand for high-end practical professionals in tourism industry and the insufficient supply of specialized talents in colleges and universities burdens the development of tourism industry. At the same time, since the 1990s, many colleges and universities in Europe and the United States have started to set up professional degree education programs to respond to the rapid development of tourism industry and the large demand for pragmatic management talents. In line with the development trend of tourism higher education worldwide and the domestic need for high-end talents with practical ability, the Academic Degrees Committee of the State Council issued the *Notice of Distributing the Setup Plans of 19 Professional Degrees including Masters of Finance*, which clarified the specific setting scheme of MTA education program. In September 2010, the MTA education program has been included in the national postgraduate enrolment plan for the 2011 academic year, marking a new stage of China's postgraduate education in tourism studies.

A professional degree is a degree type set to respond to the needs of the society for the specific professional fields, which cultivates high-level practical talents who have strong professional abilities and are able to creatively engage in practical work. A professional degree has a relatively independent education mode with a strong emphasis on career development. With the rapid development of China's tourism industry, the MTA program was approved by the Academic Degrees Committee of the State Council in March 2010 to meet the urgent needs for training tourism professionals for the development of China's tourism industry. Meanwhile, the MTA program provides an opportunity for improving the training system, innovating the training mode, and improving the training quality of tourism management talents. The MTA education program is also a part of tourism management postgraduate education in China. The establishment of the MTA education program is of great significance to the tourism management postgraduate education in China. Postgraduate education with professional degree has been put as a development priority which will be strongly supported and actively guided in the following decades. Such top-down support can effectively adjust the postgraduate structure of tourism management, promote the innovation of training mode and management mechanism, and improve the training quality of the MTA education program, which is becoming increasingly important in the cultivation of tourism practical talents.

Rapid expansion of the MTA education

In September 2010, MTA was listed in the 2011 Unified National Graduate Entrance Plan by the Ministry of Education, allowing 57 universities from 24 provinces to launch the MTA education program and enrolling students, thus marking the official launch of China's MTA education program in the whole country. On March 18, 2011, the China National MTA Education Supervisory Committee was established. It is a professional organization for graduate education in Tourism Management under the guidance of the Academic Degrees Committee of the State Council, the Ministry of Education, and the Ministry of Human Resources and Social Security. Its secretariat is affiliated with Nankai University. Since then, the MTA education program has become an important part of the tourism management postgraduate education. The tourism management postgraduate education has shifted from focusing on the cultivation of academic talents to the cultivation of both academic and applied talents.

Since the start of the MTA education enrolment, after nearly a decade of educational exploration, the MTA education system has begun to take shape. In 2011, there were 57 MTA training institutions in China; by the end of 2017, there were 97 MTA education program authorized institutions; and by the end of 2019, there were 104 MTA education and training units in China. After nearly a decade of development, the MTA education program has achieved rapid development. The cultivation of students by MTA education and training units is constantly improving in the areas of teaching modes, teaching contents, and teaching methods.

Geographic and institutional distribution

Distribution of MTA education institutions

By the end of 2019, there were 104 universities and research institutions with independent conferment and enrolment rights for master degree in tourism management in mainland China, covering 30 provinces and autonomous regions (as shown in Table 7.1). In terms of the provincial and municipal distribution of these institutions, Beijing, Liaoning Province, and Shandong Province each host six institutions, followed by Guangdong Province, Hunan Province, Henan Province, Zhejiang Province, Jiangxi Province with five each; and the least is the Tibet Autonomous Region with one. The distribution pattern basically reflects the imbalance of tourism development and tourism education in China.

Regional distribution of MTA education institutions

From the perspective of geographic distribution, 104 MTA education institutions basically cover Northeast China, Northern China, Eastern China, Central China, Southern China, and Southwest and Northwest regions (Table 7.2). Most of them are distributed in Northern China with a total of 24 institutions but the distribution is relatively concentrated, 22 in Eastern China, 13 in both

Table 7.1 Provincial and municipal distribution of recruitment units

Serial number	Provinces	Number	Serial number	Provinces	Number
1	Beijing	6	16	Henan	5
2	Tianjin	3	17	Hubei	3
3	Hebei	3	18	Hunan	5
4	Shanxi	3	19	Guangdong	5
5	Inner Mongolia	3	20	Guangxi	4
6	Shandong	6	21	Hainan	3
7	Liaoning	6	22	Chongqing	4
8	Jilin	2	23	Sichuan	4
9	Heilongjiang	2	24	Guizhou	2
10	Shanghai	4	25	Yunnan	3
11	Jiangsu	3	26	Shaanxi	3
12	Zhejiang	5	27	Gansu	2
13	Anhui	2	28	Qinghai	2
14	Fujian	3	29	Xinjiang	2
15	Jiangxi	5	30	Tibet	1

Source: China National MTA Education Supervisory Committee.

Table 7.2 Regional distribution of MTA program authorizations

Serial number	Regions	Number
1	Northern China	24
2	Eastern China	22
3	Central China	13
4	Southwest China	13
5	Southern China	12
6	Northeast China	10
7	Northwest China	10

Source: China National MTA Education Supervisory Committee.

Central China and the Southwest, 12 in Southern China, and ten each in the Northeast and Northwest regions. These institutions are mainly concentrated in Liaoning Province and Shaanxi Province.

Scales

According to the overall enrolment of students from 2011 to 2016, the number of enrolled students in MTA education programs shows a steady growth trend in China (Figure 7.1). In 2011, the actual number of registered personnel was 319; in 2012, the number was 471, an increase of 47.65% compared with 2011. In 2014, the actual number of registered students was 608, an increase of 7.23% compared with 2013. In 2015, the total number of registered students was 659, an increase of 8.39% compared with 2014; in 2016, the total number of registered students was 716, an increase of 8.65% compared with 2015 (Bai &

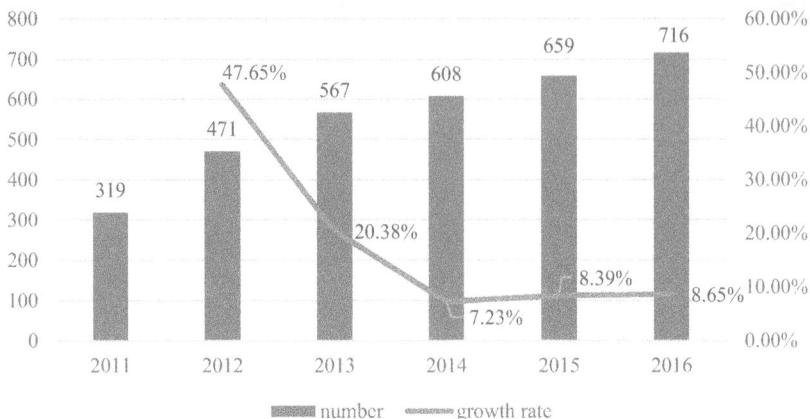

Figure 7.1 The actual number of enrolled Masters of Tourism Administration from 2011 to 2016.
Source: Bai & Tuo (2018), p. 90.

Tuo, 2018). Universities and scientific research institutions that have the right to grant the degree of MTA and enrol students are also increasing the enrolment publicity of the program by constantly improving the capabilities of teachers, innovating teaching methods with the help of MTA case development and other projects, improving the quality of teaching, and attracting more students to apply for MTA education program.

Operational modes

Study mode

MTA education in China implements a dual track system of full- and part-time on-the-job and off-job study modes (Liu & Li, 2016). The learning process of MTA education program highlights the practical guidance of tourism management and strengthens practice-based teaching. The practice teaching time is no less than half a year. In terms of training mode, the MTA education program pays attention to the close cooperation between schools and the relevant departments and agencies of tourism management practice to jointly discuss, formulate and implement the content of the training program. Furthermore, MTA education program attaches great importance to engaging industry professionals to undertake the major course teaching and constructs a "dual qualified teacher" structure, under which each student is equipped with tutors from both the academia and the industry. The tutor team is responsible for the daily management of postgraduates, academic and moral education, formulation and adjustment of postgraduate training plan, guidance of literature reading, courses selection and participation in practice and innovation activities, organization and arrangement, guidance of scientific research and graduation/dissertation, and so on. For example, the MTA program of Nankai University can be divided into part-time and full-time modes. Full-time and part-time students have different class times, teaching courses and training methods. It reflects different training characteristics and training objectives and implements the training mode under the guidance of double tutor system (MTA Education Centre of Nankai University).

Research direction

Each university with the authorization of MTA education program has set up different research directions according to its own actual situation, which, to a certain extent, reflects the school's strength, characteristics and advantages. It lays the foundation for the development of the MTA education program and also adapts to the needs of various tourism talents in social practice. All over the country, according to the research characteristics and advantages of each authorized institution, there are more than ten research directions under the MTA education program. The major direction of the MTA education program mainly focuses on tourism enterprise management, hotel management, attraction management, tourism planning and development, tourism destination management, and so on; but at present, there are a few emerging research directions. There are also schools relying on the advantages of disciplines and teachers, setting up a cutting-edge research direction. For example, relying on the advantages of management discipline and tourism management education, Nankai University has set up research directions such as "big data and tourism management", "tourism innovation and entrepreneurship", "new tourism formats", and so on; Dongbei University of Finance and Economics set up the direction of "leisure and service management" based on the characteristics of tourism experience research.

Training objectives

According to the "guidelines of training program for postgraduate master degree in Tourism Management (Trial)" issued by the MTA Education Supervisory Committee, the goal of the MTA education program is to cultivate high-level and applied talents with social responsibility, entrepreneurship and innovation ability, professional quality, international perspective and strategic thinking, and capable of practical work in modern tourism and related industries. In the training process of the MTA education program more attention is paid to the combination of theory and practice, knowledge and ability so as to improve students' ability to analyse and solve problems and their comprehensive quality. According to their respective training orientation, each MTA educational authorization unit has refined the training objectives and formed its own unique training objectives. For example, the training objective of the MTA education program at Nankai University is to master the solid basic theory and broad professional knowledge in the field of tourism management, and have a sense of social responsibility, entrepreneurship and innovation ability and tourism occupation accomplishment, have an international perspective and strategic thinking ability, and cultivate high-level applied talents capable of practical work in modern tourism and related industries.

Length of study

In China, the length of study for the MTA education program is generally two to three years. Credit system and flexible education system are implemented, meaning that students can apply for early or delayed graduation. During the basic study years, the first year or 1.5 years will be used for course study and the last year or 1.5 years will be used for social practice, social investigation, writing graduation thesis and thesis defence. For example, the length of schooling of Nankai University's MTA education program is 2.5 years, with a maximum of three years. The first 1.5 years will be for centralized class teaching and the next year for practical teaching and graduation thesis. The length of schooling of the MTA education program at Sun Yat-sen University is three years and the study mode offered is part time (the teaching time is usually arranged on weekends).

Curriculum structure

Curriculum setting

At present, the MTA education program curriculum is composed of core compulsory courses and directional elective courses. The teaching goal of the core compulsory courses is to train students to master the knowledge and skills of tourism management, and to cultivate adaptability and creativity to engage in tourism management. The core compulsory courses lay the foundation of professional knowledge, ability and qualities that students should possess. The elective courses aim at providing a wide range of knowledge and a complete knowledge structure for students, which are mainly composed of courses related to tourism, with teaching goals to lay a broader professional foundation for the training of students and enhance the flexibility of talent training.

Curriculum content

The core compulsory courses and elective courses for the MTA education program should be able to provide students with the necessary knowledge and skills to work in the tourism industry. According to the "Guidelines of Training Program for Postgraduate Master Degree in Tourism Management (Trial)" issued by the China National MTA Education Supervisory Committee, the curriculum is implemented with a credit system with no less than 32 credits. Among them, there are nine core compulsory courses, including Marxist Economic Theory, Professional English, Tourism Industry Economic Analysis, Tourism Destination Development and Management, Tourism Marketing, Tourism Planning and

Strategy, Tourism Investment and Financial Management, Service Management, and Tourism Information System. Each training unit can set up elective courses around the professional direction according to students' characteristics, teaching advantages and talent training features. It is recommended that each elective course should be between 1 and 3 credits. The elective courses are mainly set up to expand students' knowledge and refine the courses of tourism management, such as Tourism Consumer Behaviour, Tourism Policies and Regulations, Tourism Culture, Tourism Project Investment and so on.

As tourism management is a comprehensive area of specialization, most of the universities set up courses differently based on theories of management and economics. However, in China, the professional elective courses for the MTA education program vary greatly with research directions and disciplines they rely on. Colleges and universities offer elective courses according to their own discipline advantages and expertise of teaching staff.

Teaching methods

The professional master degree program in tourism management is different in training objectives and teaching methods from that of academic master degree program. MTA education program aims at enlightening students' thinking and cultivating students' ability to analyse and solve problems by multiple teaching methods such as course teaching, case study, teamwork, and professional internship. The training process highlights the practical orientation of tourism management. The practice-based learning is no less than half a year.

Case teaching is important in MTA education. In order to constantly promote the reform of MTA teaching methods, the training unit has carried out the innovation of teaching methods based on industry practice, which adopted four teaching forms: classroom case teaching, industry elite lectures, field trips, and practical research around special themes. For example, the MTA education program of Nankai University is committed to providing students with international learning perspective and experience. It adopts international curriculum design and combines heuristic and discussion teaching methods. Combining theory with practice and paying attention to practical application, Nankai University is devoted to cultivating students' ability to analyse and solve problems.

In addition, teaching methods are based on the actual needs of the development of China's tourism industry, and classroom discussions are fully used to guide students to think creatively in conjunction with their work. Specifically, the teaching methods include special thematic lectures, business investigations, classroom case discussion, case writing, business simulation training, practical research and investigation, domestic and overseas internships, international exchange, and domestic and overseas study tours. It is supposed to train students to become tourism industry leaders with an international perspective, rational decision-making ability, leadership and social responsibility (MTA Education Centre of Nankai University).

Course assessment

The course learning requirement of the MTA education program is to complete the corresponding credits of core compulsory courses and elective courses. At the end of each course, students have to participate in course examination, and those who pass the examination are awarded the course credits. In terms of the MTA education program curriculum assessment, the "Guidelines of MTA Training Program" stipulates that students' academic performance is comprehensively evaluated in terms of exams (including oral exams), assignments, classroom discussions, case analysis, special reports, and literature reading.

Assessment

Academic thesis

According to the requirements of the *MTA Degree Setup Plan*, students' thesis topics should consider the actual development of the tourism industry. The thesis can be in the form of a specialised study, a high-quality survey report, or a case study report. Academic thesis must integrate theory into tourism practice, extracting scientific problems from specific tourism practices and closely focusing on the problems that need to be solved in the development practices of tourism and related industries. The research results are expected to have application value. This reflects students' ability to use tourism management and related subject theories, knowledge, and methods to analyse and solve practical problems in tourism management.

The thesis should generally have a literature review section, stating how existing theories and studies are related to the practical problem with an emphasis on explaining and introducing the main theories used in the thesis. The thesis should have a clear research design, explain the process of data collection and the source of data. The analysis and evaluation of data should be scientific and reasonable, reflecting the relationship with existing theories. Thesis using survey methods, case study, and field research methods must present valuable and innovative real-world problems and reflect the value of the research questions with abundant first-hand data. Research materials cannot be simply accumulated in writing. Instead, they should be systematically identified, synthesized, organized, and analysed following the learned theories. Based on data analysis, students are supposed to draw their own conclusions.

Thesis evaluation mainly involves assessing the student's ability to use the theories learned to solve practical problems, whether the content has new insights, and its practical value (such as contribution to social and economic benefits). The thesis is evaluated by a combination of internal and external examiners with varied forms of thesis defence. The members of the thesis steering group must include senior managers in tourism or related industries with rich practical management experiences and achievements. Among the thesis examiners, there must be an expert holding a senior professional position in the tourism industry. The thesis defence committee must have experts who work in the tourism industry and hold senior professional and technical positions or senior management positions.

Graduation and degree award

In general, students who have completed the training courses, such as course study and internship practice within the prescribed study period, earned the prescribed credits, and met the other requirements for graduation, will be allowed to graduate and will be issued a graduation certificate. Moreover, students who meet the requirements for degree application will be awarded the master's degree in tourism management by the degree evaluation committee of the degree-granting unit.

Problems and issues

Some educational courses lag behind

The tourism industry in China has been experiencing rapid development, requiring that the content of the MTA education program curriculum keep pace with the changes in the industry, and timely grasp of the industry dynamics and industry development directions on hot issues. Thus, the MTA curriculum needs to be professionally kept up-to-date. The course content is an effective carrier to achieve teaching objectives. The rapid development of tourism has led to the continuous improvement of tourism education. However, the contradiction between the speed of knowledge updating and the lagging in the cultivation of high-level tourism talents has become increasingly prominent. Therefore, the course content must be advanced, solid and comprehensive.

In China, the content of the MTA education program curriculum generally lacks foresight, which leads to a lack of attention to the hot issues, to the problems and directions of industrial practice, and to the difficulties of tourism research. For example, there is a lack of corresponding courses and lectures for online travel companies, leisure travel, elderly travel, health and wellness travel, and sports travel.

Lack of special teaching materials

The Peking University Tourism Planning Research Centre conducted a survey on tourism textbooks. The results showed that there are only 27 textbooks for graduate students in tourism, accounting for only 5% of the total number of tourism textbooks (Wu & Li, 2006). This is obviously not conducive to the cultivation of graduate students majoring in tourism management. Moreover, the textbooks required for professional graduate students are different from academic graduate students, and their textbooks should pay more attention to the management issues in the industry. On the one hand, the theoretical knowledge in the textbook can guide them to solve management problems in the real industry; on the other hand, industry management case textbooks as well as the excellent and classic management cases can improve students' ability to analyse and solve problems. Up to now, lecture is the main teaching method, resulting in rare interactions with students. Without group discussions and field investigations, the teacher-centred teaching method neglects students' active role in the learning process, leading to poor student participation and involvement.

Teachers' capabilities need to be improved

Teachers are the fundamental guarantee for the quality of the MTA education program. There is a shortage of "double-qualified" teachers in the MTA education program. "Double qualified" teachers are compound talents with both teaching abilities and corporate work experiences. The abilities of teachers cannot fully meet the needs of the MTA education program and can't meet the needs of talent training for the tourism industry. On the one hand, in the current teaching workforce, there are many teachers who did not graduate from a tourism management major or lack working experiences in the tourism industry; as a result, they cannot grasp the industry dynamics in time and their knowledge and skills lack pertinence to the tourism industry. On the other hand, although many universities hire leaders of tourism industry and officials of tourism administrations as part-time professors or industry mentors, these part-time professors or mentors teach mainly in lectures and give limited guidance to students. Compared with the standards of teaching staff in foreign well-known tourism universities, the qualifications of teaching staff in China's MTA education program still needs to be improved.

Conclusion

This chapter introduces the background, development routes and enrolment size of the MTA program in China. It also introduces the general situation of MTA education program in four aspects: operational modes, curriculum structure, assessment, and problems and issues. China's MTA education program has its own characteristics, such as diversified school-running models, gradual improvement of the curriculum system and professional settings, the teaching model of applying theory into practice, diversified student evaluation systems, innovative mode of teaching, and the interactive development of internationalization and localization. China's MTA education program can be improved by updating the curriculum content, enhancing teacher's strength, and compiling high-quality teaching materials.

References

Bai, C. H. & Tuo, Y. Z. (2018). Annual development report of Education Program for Master of Tourism Management (MTA) in China (2016–2017). In Tourism Education Branch of China Tourism Association (Ed.), *China Tourism Education Blue Book 2017–2018* (pp. 88–105). Beijing: China Tourism Press.

Liu, J. & Li, Y. (2016). A probe into the realistic thinking and countermeasures of the development of postgraduate education for master's degree of tourism management in China. *Degree and Postgraduate Education*, 10, 16–21.

Wu, B. H. & Li, X. X. (2006). Report on the development of tourism education in China. *Tourism Tribune*, 21(S), 9–15.

8 Tourism PhD programs in China

Honggang Xu and Qingfang Zhang

Introduction

PhD programs are designed primarily for high-level talent capable of systematic and advanced research work, and are typically offered only by research-oriented educational institutions such as universities (UNESCO Institute for Statistics 2012). Since Nankai University took the lead in recruiting PhD students in tourism geography in 1986, the tourism PhD education in China has gone through over 30 years' development, with programs evolving in both quality and quantity. Currently, tourism PhD education is not only offered in geography programs, but also in other disciplines such as sociology and anthropology. PhD programs in tourism management, as independent programs, are also growing. These tourism programs contribute significantly to academic research, higher education and policy-related and industrial development in the industry.

This chapter reviews the historical development and evolution of tourism higher education. Specifically, the characteristics of PhD programs, dissertation topics and quality assurance are examined. Potential issues and challenges are also discussed.

Development of tourism PhD education in China

In China, disciplines that can offer PhD degrees are strictly managed and must be listed in subject categories, issued by the Office of State Council Academic Degree Committee (SCADC). The China SCADC subject categories stipulate the disciplines that can confer academic degrees, first-level disciplines and secondary disciplines of talent training in a hierarchical order.

According to the *Subject Catalogue of Degree Awarding and Talent Training* (updated in 2018), 13 fields have been included: philosophy, economics, law, education, literature, history, natural science, engineering, agriculture, medicine, military science, management science and art. The first-level discipline is the subject category, and within each field, there are many categories. The secondary discipline is the subcategory/specialty within the corresponding first-level discipline. At present, institutions of higher learning can recruit PhD students in related first- and second-level disciplines. Since each first-level discipline includes

DOI: 10.4324/9781003004363-8

several secondary disciplines, when degree-conferring institutions are authorised to grant doctoral degrees in first-level disciplines, all secondary disciplines under this category will automatically be granted the same right. However, in certain circumstance, a secondary-level discipline can also offer doctoral degree program if permitted to do so by the Provincial Education Department.[1]

Historical development of tourism PhD education

Doctoral education in tourism geography

Doctoral-level tourism education in China began with geography, which was recognised as a well-established first-level discipline. In the 1980s, with the painstaking efforts of older-generation geographers represented by Chen Chuankang and Guo Laixi, tourism geography was recognised as a sub-subject in geography. In 1986, Nankai University was the first to recruit PhD candidates in tourism geography, marking the beginning of higher tourism education in China. Up to 2000, only five universities, Nankai University, Peking University, Sun Yatsen University, Nanjing University and East China Normal University, provided PhD programs in tourism geography. In addition, the Institute of Geographic Sciences and Natural Resources Research of the Chinese Academy of Science (CAS) also undertook the task of cultivating doctoral students in tourism geography (Bao 2002; Chen 2004b).

During this period, experience was gained in doctoral tourism education. Due to strict control by the Chinese government over the qualifications of doctoral supervisors, only about six professors in China were qualified to supervise doctoral students in tourism geography, and the number of tourism PhD students was limited (Bao 2002). From 1989 to 2000, a total of 20 students from the above six institutions successfully obtained a doctorate degree in tourism geography (Bao 2002). By comparison, in North America, there were 76 doctoral dissertations in tourism studies in geography from 1951 to 2000 (Meyer-Arendt 2000; Meyer-Arendt & Justice 2002).

Doctoral education in tourism management

In 1997, the SCADC and the State Education Commission of the People's Republic of China (PRC)[2] jointly issued the *Discipline and Subject Catalogue of Doctor and Master Degree Awarding and Graduate Training* (promulgated in 1997). In this document, management science was approved as an independent discipline, and under this field, tourism management was recognised as a secondary discipline within the first-level category of science of business administration. As a result, any university that had obtained approval from the Ministry of Education of the PRC to offer a PhD program in science of business administration could launch a tourism management program and offer PhD programs in tourism management. Thus, the postgraduate education of tourism majors entered a stage of development. In 2000, Sun Yat-sen University,

Fudan University, Xiamen University and Dongbei University of Finance and Economics became the first four universities in China to offer doctoral degrees in tourism management. In the same year, Sun Yat-sen University began to recruit PhD students, marking the official start of tourism management PhD education in China.

Universities that could not offer PhD programs in the first-level discipline of science of business administration could apply to the Ministry of Education of the PRC to offer PhD programs for tourism management only. From 2003 to 2005, 11 universities and institutions started tourism management PhD programs: Renmin University of China, Nankai University, Zhejiang University, Xiamen University, Shanghai University of Finance and Economics, Jinan University, Shaanxi Normal University, Zhongnan University of Economics and Law, Yunnan University, Fudan University and the Institute of Chinese Academy of Social Sciences. By 2020, 64 institutions of higher learning and one research institute (the Institute of Applied Economics, Shanghai Academy of Social Sciences) had PhD enrolment in tourism management and related research fields. According to statistics released by the Ministry of Culture and Tourism of the PRC, enrolment of PhD students in tourism and tourism-related majors in China had expanded from 167 in 2014 to 336 in 2017.

Distribution of tourism PhD programs in different disciplines

Due to institutional reform, the Ministry of Culture and Tourism of the PRC stopped publishing official statistics on tourism education (including PhD education in tourism) in 2018. In this chapter, institutions of higher learning that recruit doctoral students in tourism-related areas under a first-level discipline and in tourism-related research areas in certain second-level disciplines (including tourism management) are regarded as higher education institutions that offer doctoral degrees in tourism.

To identify institutions of higher learning that provide doctoral degrees in tourism, we reviewed the general rules and professional directories for PhD enrolment published by all 827 public institutions of higher learning at undergraduate levels or above in the *List of National Institutions of Higher Learning in 2020* released by the Ministry of Education. Sixty-four institutions of higher learning offering tourism PhD degrees were identified, and a further analysis of these 64 institutions was conducted in order to grasp the discipline backgrounds and distribution of colleges and universities in China's doctoral tourism education.

Among the 13 fields of discipline recognised in the *Subject Catalogue of Degree Awarding and Talent Training* (updated in 2018), a total of nine disciplines could offer doctoral education in tourism. PhD students in tourism can graduate with a degree from fields such as management science, natural science, engineering, agriculture, education, law, economics, history and philosophy. As shown in Figure 8.1, management science (62%) and natural science (17%) are the main fields that award tourism PhD degrees.

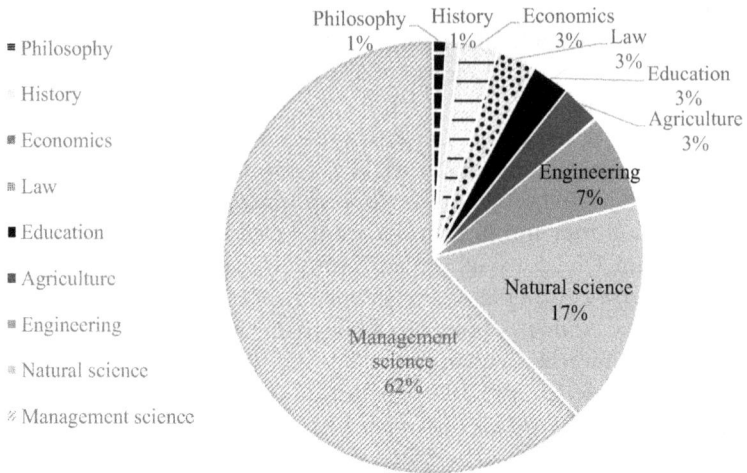

Figure 8.1 Distribution of discipline fields for tourism education.
Note: Data were compiled by the authors.

A further analysis of disciplines that offer doctoral degrees in tourism was carried out. As shown in Table 8.1, eight first-level disciplines could offer doctoral degrees in tourism, including philosophy, Chinese history, geography, environmental science and engineering, landscape architecture, forestry, management science and engineering and science of business administration. In addition, 11 secondary disciplines, including land resources and ecological economics, sociology, industrial society and management, human and sociological science of sports, human geography, landscape ecological planning and management, tourist geology and geological relics, tourist management, corporate management and environmental management and marketing, also offer doctoral degrees in tourism. Among them, tourism management (a secondary discipline), business administration (a first-level discipline) and human geography (a secondary discipline) are the top three disciplines for doctoral tourism education. There are 30 institutions offering PhD degrees in tourism management, 13 of which fall under the discipline of business administration and ten institutions fall under human geography.

In addition, each institution has its own research focus (see Table 8.1). These include tourism planning and development, tourism economic development, tourism sustainable development, enterprise management, destination management, rural tourism, leisure tourism, MICE (meetings, incentives, conferencing and exhibitions) tourism, tourism for older people and tourism big data.

Distribution of institutions offering tourism PhD programs

In 2020, 64 institutions of higher learning enrolled PhD candidates in tourism management-related research directions. Among them, 42 institutions of higher

Table 8.1 Subordinate disciplines for recruiting tourism PhDs and the distribution of tourism colleges and universities

Field of discipline	The first-level discipline	Secondary discipline/major	Number of colleges and universities	Research fields
Philosophy	Philosophy	–	1	Tourism ethics; environmental ethics; tourism philosophy
Economics	Applied economics	Land resources and ecological economics	1	Tourism industry economics; tourism finance; ecological resources and tourism development
Law	Sociology	Sociology	1	Cultural heritage; folk culture; tourism research
		Industrial society and management	1	Post-industrial society and tourism management; tourism management
Education	Science of physical culture and sports	Human and sociological science of sports	2	Leisure and sports tourism; sports tourism; tourism management
History	Chinese history	–	1	Cultural history of tourism
Natural science	Geography	Human geography	10	Tourism geography and tourism planning; tourism development and planning; tourism geography and resource development and utilisation; urban leisure and tourism planning; tourism functional area planning and design; tourism development and management; regional tourism management and planning; regional development and tourism planning; regional tourism development and management; tourism ecology and environmental studies; Karst cultural geography and tourism; tourism and cultural geography; historical geography and tourism management; social geography of tourism
		Landscape ecological planning and management	1	Ecological tourism and scenic spot planning
		–	2	Tourism regional effect and tourism planning; natural and cultural tourism landscape research; tourism ecology and environment research; tourism planning and management research

(Continued)

Field of discipline	The first-level discipline	Secondary discipline/major	Number of colleges and universities	Research fields
Engineering	Geological resources and geological engineering	Tourist geology and geological relics	1	Tourist geology and geological relics
	Environmental science and engineering	–	1	Tourism planning and environmental management
	Landscape architecture	–	1	Tourism planning; nature reserve and recreation management
Agriculture	Forestry	–	2	Forest recreation and park management
Management science	Management science and engineering	–	3	Agricultural systems engineering and tourism management; sustainable development and tourism management; hotel and tourism management
	Science of business administration	Tourist management	30	National tourism development and management; tourism development and management and regional economic strategy; tourism development and enterprise management; regional tourism and tourism planning; tourism planning and smart tourism; tourism development and landscape conservation; tourism industry operation management; tourism development management and information technology; tourism development and management research; tourism planning and resources development; heritage resources development; regional tourism development and planning; tourism enterprise management; service management; service innovation and strategic management; service management; tourism and modern service industry; post-industrial society and tourism management; tourism development and market analysis; tourism marketing; brand management; tourism destination competitiveness and supply chain management; tourism experience; leisure travel consumer behaviour and marketing management; tourism theory and market; tourism marketing and consumer behaviour; tourist marketing and tourist behaviour research; tourism marketing business intelligence analysis and decision; service management and marketing; tourism destination research; tourism destination management and marketing; tourism industry and regional development; tourism and regional development research; tourism economy and industrial management; tourism industry economy; tourism collaborative innovation; tourism economy and management innovation; tourism industry management; tourism

		industry ecology; tourism engineering research; tourism economic operation and crisis management; tourism development evaluation and management; tourism economic management; tourism enterprises and tourism industry research, science of tourism management and decision-making, research on tourism policy and development, tourism sustainable development, research on the sustainable development of mountain tourism; the integration of new forms of cultural tourism; research on the transformation and sustainable development of tourism industry; industrial reconstruction and decision-making under the influence of major crisis events; research on tourism and leisure development; leisure industry and tourism economy; leisure social psychology; leisure agriculture and rural tourism development; rural experience tourism and community management; tourism culture research; cultural tourism and regional development; cultural heritage development and management; cultural tourism industry development and management innovation; cultural tourism integration; festival tourism and exhibition management; tourism resource management; elderly tourism; health care tourism; leisure accommodation industry and health care management; old-age care services; tourism education innovation and big data research; tourism case-based reasoning and data mining; tourism sociology; tourism mobility and tourism transportation
Corporate management	3	Tourism planning, tourism management
Environmental management	1	Tourism destination management; community tourism and community participation; ecological security evaluation
Marketing	1	MICE (meetings, incentives, conferencing and exhibitions) tourism management
–	13	Tourism management; tourism service and management innovation; yacht management; tourism resource development and management; tourism strategic planning; tourism planning and development; tourism planning and tourism scenic spot management system research; tourism and regional economic development; tourism economy; tourism and community development; event management and tourism development; tourism economic management; marketing, tourism marketing, and service management; tourism development and destination management; tourism big data; tourism e-commerce; tourism enterprise strategic management; exhibition and event management

Data source: The author collated the data according to the general rules and professional directories of PhD enrolment published by 64 universities in 2018–2020.

Notes: '–' refers to the situation that students are directly recruited in the first-level disciplines; the same institute of higher learning may recruit tourism PhDs in different disciplines or majors.

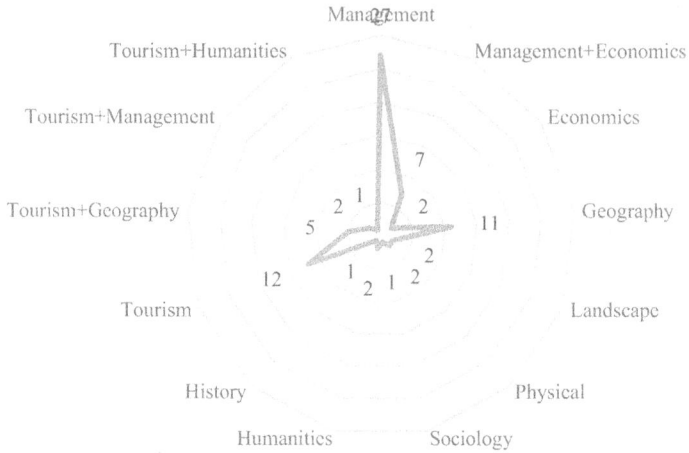

Figure 8.2 Types of departments offering PhD education in tourism and related research areas.

Data source: The author collated the data according to the general rules and professional directories of PhD enrolment published by 64 universities during 2018–2020.

Note: Some schools offer doctoral education in tourism and related research directions in different departments.

learning were in national key construction projects (21 universities with Project 985 and 21 universities with Project 211).

A summary of 64 domestic colleges and universities that offer tourism PhD degrees (Figure 8.2) shows that independent management-related departments (27), tourism-related departments (12) and geography-related departments (11) are the main departments in colleges and universities that provide tourism PhD programs. For example, Sun Yat-sen University has an independent tourism school. In addition, 12 independent tourism departments and eight institutions of higher learning have established tourism-related departments jointly with other subjects. Among these departments, five were founded jointly with geography subjects, two with management subjects and one with humanities subjects (see Figure 8.2). This further verifies that geography and science of business administration are two important disciplines that cultivate tourism PhD students. Tourism management has thus become an important secondary discipline in business administration.

Basic information of tourism PhD programs

To have a clear understanding of the general situation regarding doctoral tourism programs, we summarised the training programs published by 16 higher education institutions with an authorised independent discipline of tourism management (Beijing Jiaotong University, Capital University of Economics and

Business, Xiamen University, Sun Yat-sen University, Jinan University, Zhongnan University of Economics and Law, Dongbei University of Finance and Economics, Ocean University of China, Northwest University, Fudan University, East China Normal University, Southwestern University of Finance and Economics, Nankai University, Yunnan University, Yunnan University of Finance and Economics and Zhejiang University).

Basic structure of tourism PhD programs

The overall aim of doctoral tourism education is to cultivate theoretical research capability and provide students with a solid professional foundation, international academic vision and innovative spirit as well as to provide senior management talent in enterprises and government departments.

Most PhD programs take three years to complete, but some universities, such as Sun Yat-sen University, have begun to offer four-year programs. The programs often combine course learning and dissertations, including one year of coursework and two to three years of dissertation work. Universities have increasingly requested that students publish journal papers to obtain a degree.

Basically, PhD students are well funded. Almost all students hold scholarships, apart from some international students. PhD students are encouraged to attend academic conferences, and universities also create favourable conditions to fund doctoral students to visit other famous universities and research institutes at home and abroad for study and research.

Course system in tourism PhD education

Based on 16 higher education institutions with authorised independent programs of tourism management, the degree course system consists of four levels: public degree courses, basic degree courses, specialised degree courses and specialised elective courses. Among them, public degree courses, basic degree courses and specialised degree courses are required.

Public degree programs include courses in political theory and a first foreign language, which is regulated by the Ministry of Education of PRC. The political theory course, as an important part of China's postgraduate education, aims to strengthen the moral education of postgraduate students and improve their ideological, political and moral qualities, while a foreign language is an essential quality of high-level talent and a critical requirement for functioning in a time of economic globalisation and the increasing internationalisation of academic research networks.

The basic degree courses are mainly related to management, such as advanced microeconomics, Western economics, advanced econometrics, management research methods, advanced management, theory and practice of business administration and other related management and economic courses.

Specialised degree courses are intended to expand on the basic theory and to master systematic professional knowledge. When based on tourism management, they reflect the characteristics of the secondary discipline and highlight professionalism. These courses include tourism industrial economics, destination planning theory and management, tourism management theory, research on the philosophy of tourism, cultural tourism and so on. Both the basic and the specialised degree courses attach great importance to becoming familiar with cutting-edge issues as well as theory and literature required for research projects.

Specialised elective courses are mainly designed to further broaden students' basic theories and their scope of knowledge. Most universities in China set elective courses for tourism majors in accordance with the research field of supervisors. In order to encourage postgraduate students to take the initiative in pushing forward certain academic issues and to track academic progress, some colleges and universities also offer academic lectures and discussions in addition to training courses and workshops for grasping specific knowledge in the field of tourism, academic reading and writing skills and so on.

Generally speaking, 20 credits are required for doctoral education with a tourism major. However, in most universities in China, there is a disproportionate number of required courses and elective courses in the doctoral programs in tourism. Usually, there are two to four specialised required courses and one to two elective courses. In addition, course credits are also uneven, with required courses usually having a high score of three to four credits, while elective courses usually have one to two credits. Notably, the overemphasis on compulsory courses is not conducive to individualised teaching or stimulating students' active learning ability. Further, there are few cross-disciplinary courses in the existing doctoral programs in tourism, which limits the broadening of students' scope of knowledge.

Quality control of dissertations

Among all the required learning components, the doctoral dissertation is an important factor in obtaining a PhD. Tourism colleges and universities all have strict requirements for a doctoral dissertation: (1) the dissertation should have creative research results that are of theoretical significance and practical value in the research field; (2) the dissertation must be based on a systematic and complete academic project with a broad scope and depth; and (3) the dissertation should present sound conclusions with clear viewpoints, sufficient arguments, strong logic and a rigorous structure. If plagiarism is found in the dissertation, the dissertation defence will be cancelled.

In order to improve the quality of doctoral dissertations, universities require that the writing process be carried out in strict accordance with certain procedures. Universities mainly conduct quality control of doctoral dissertations through procedures for thesis selection, proposal presentation, midterm assessments, pre-defence, anonymous reviews and formal oral defence. The specific time of the proposal defence in the candidature is mainly decided by the supervisor or the department, but the time interval from the confirmation of the proposal to the application for oral defence of the dissertation is generally no less than one year. In the midterm assessment, a team of three to five faculty members within the scope of second-level disciplines convene to assess the comprehensive ability, progress of thesis work, working attitude and workload of the candidate. Only after passing the midterm assessment can the candidate proceed to the next stage of work. Doctoral students can apply for oral defence after completing their personal training plan, meeting the training plan for their discipline and receiving external examiner approval.

Whether the defence is successful or not will be decided by the defence committee through anonymous voting, and a majority affirmative vote (generally two-thirds) means the defence has passed. The defence committee is generally composed of five or more members (excluding the supervisor), and dominated by experts from the school, with approximately two experts from outside the school.

Knowledge contributions of tourism doctoral dissertations

Doctoral dissertations are important academic works that show the latest frontier knowledge and research achievements, including in the area of tourism (Huang 2011; Weiler, Moyle & McLennan 2012; Ying & Xiao 2012). Thus, an analysis of doctoral tourism dissertations was carried out to understand the trends over the years regarding the number of doctoral dissertations, the discipline backgrounds and the topics.

The China Doctoral Dissertations Full-Text Database (CDFD, http://www. wanfangdata.com.cn) on the Wanfang Data Knowledge Service Platform was used to conduct the search. The literature retrieval criteria were 'Subject = Tourism' or 'Title or Keyword = Tourism/Hotel/Restaurant/Tour operator/Convention and exhibition'; a total of 2,901 articles were retrieved. After manually deleting articles that were irrelevant to tourism and eliminating duplicated cases, a total of 1,395 tourism dissertations (1989–2020) were collected (available on June 10, 2021). Due to the late establishment of the CDFD, the doctoral dissertations on Tourism Geography (1989–2000) summarised in Bao's (2002) paper were supplemented, and thus a total of 1,399 doctoral dissertations on tourism were obtained.

Growth of doctoral dissertations on tourism

After Nankai University recruited the first PhD candidate in tourism geography in China in 1986, the candidate's dissertation *Study on Spatial Organization of Tourism*, produced in 1989, opened the door for doctoral dissertations on tourism. Subsequently, China's doctoral tourism dissertations entered a slow growth stage. In 1991, only two candidates under the secondary discipline of Architectural Design and Theory of Tsinghua University focused on the development of the tourism environment of the seaside resort and the reconstruction of the traditional cultural tourism pedestrian zone. Three years later, the second doctoral tourism dissertation appeared. It would be another three years before a fourth doctoral thesis on the topic of tourism appeared. Due to the small scale of training, there was little change in the number of dissertations each year. Until 2001, the annual number of doctoral dissertations on tourism was rarely greater than single digits, averaging fewer than five a year. Later, tourism dissertations in China underwent a period of rapid growth. With increases in the training, the number of dissertations increased from year to year, reflecting the urgent need for relevant research with the rapid development of tourism. After 2008, relatively stable training methods and reliable training approaches were gradually implemented, and the cultivation scale remained stable.

To be clear, the decline in tourism dissertations since 2011 in China is not due to a decline in the tourism PhD training scale. This is mainly because the country and its colleges and universities began to attach importance to the intellectual copyrights of doctoral dissertations, and more and more institutions chose to include doctoral dissertations in their libraries as unpublished articles after students' graduation, enabling university students to borrow them from their school libraries. In addition, doctoral dissertations are also available online through the university's own doctoral dissertation database instead of being published on other platforms. The public can obtain relevant information directly from a university library's official website, but the full text is usually not available. Doctoral students can also choose to publish their doctoral dissertations on an open literature service platform (usually CNKI database and Wanfang database) as long as they obtain a license for online publishing. For the protection of their own research outcomes and the intellectual copyrights of their doctoral dissertations, most tourism PhD students choose not to be included on open literature service platforms. Therefore, the online doctoral dissertations from CDFD show a decreasing trend year by year from 2010, and in 2020, there were only 18 online doctoral dissertations (see Figure 8.3).

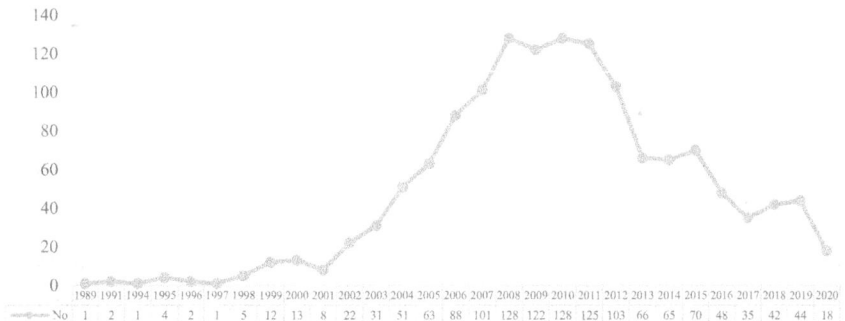

Figure 8.3 The changing trend of yearly numbers of doctoral dissertations on tourism.
Data source: The author collated the data from CDFD on the Wanfang Data Knowledge Service Platform.

Discipline distribution of doctoral dissertations in tourism studies

With the rapid growth of the number of doctoral dissertations and the development of PhD programs, the number of disciplines involved in high-level tourism research and talent training has been increasing, and the discipline composition has also undergone significant changes. Currently, in the classification of disciplines in the *Subject Catalogue of Degree Awarding and Talent Training* (updated in 2018), doctoral dissertations on tourism research involves 12 fields of discipline, 44 first-level disciplines and more than 100 second-level disciplines and research interests.

In terms of disciplines, tourism research is mainly concentrated in management science and natural science. The changes in the specific composition of disciplines with years are shown in Figure 8.4. In general, before 2000, doctoral dissertations on tourism were mainly in the field of natural science. Since 2000, the number of dissertations in the field of management science has increased sharply, which has formed two important positions focusing on the topic of tourism with that of natural science. The number of dissertations in both fields had a relatively stable trend of change, and began to fluctuate and decline after a period of an upward trend. The number of dissertations in the field of law has also shown a fluctuating upward trend since 1999, when it entered the doctoral study of tourism, to 2012, while the tourism dissertations in other disciplines have shown a large fluctuation.

In addition, the disciplines focusing on tourism research have gradually increased with the years. In the beginning of tourism doctoral education in 1989, the research topics were mainly in the field of natural science, especially in the

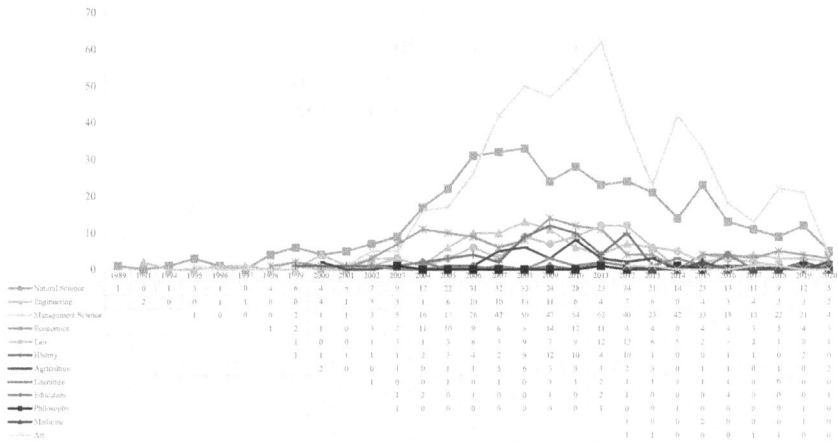

Figure 8.4 Disciplines with doctoral dissertations on tourism-related research.
Data source: The author collated the data from CDFD on the Wanfang Data Knowledge Service Platform.
Note: Some of the collected dissertations on the Wanfang Data Knowledge Service Platform have incomplete information regarding the discipline backgrounds.

secondary discipline of human geography under the first-level discipline of geography, and the main research directions included economic geography, tourism geography and tourism planning. In 1991, engineering began to pay attention to the topic of tourism, mainly in the second-level discipline of urban planning and design under the first-level discipline of architecture. In 1995, the field of management science also entered tourism research, which was initially in a first-level discipline of management science and engineering. From 1998 to 2003, PhD students in the fields of economics, law (mainly in first-level discipline of ethnology, sociology and political science), history (mainly in secondary discipline of cultural heritage and tourism development, modern and contemporary Chinese history and the history of particular subjects), agriculture (mainly in first-level discipline of forestry), literature, education (mainly in first-level discipline of science of physical culture and sports and psychology) and philosophy began to get involved in the study of tourism. This can be attributed to the rise of China's new operational types of tourism, such as ethnic tourism, tours of revolutionary heritage, agricultural tourism, ecotourism, literary tourism, sports tourism and more. In 2012, PhD-level tourism research emerged in the fields of medicine and art (mainly in the first-level discipline of design science). However, the corresponding number of doctoral dissertations is relatively small, with four dissertations, respectively, until 2020, although articles on tourism medical and health knowledge and tourism landscape design were published in Chinese journals prior to 2012.

Most first-level disciplines of various fields have focused on the topic of tourism. However, in addition to the two first-level disciplines in the field of economics, which generally is involved in tourism research, other disciplines have their own emphasis. First-level disciplines of philosophy and art do not pay much attention to tourism. Figure 8.5 indicates that among the 44 first-level disciplines involved in the doctoral dissertations on tourism research, the greatest number of dissertations was in the first-level disciplines of science of business administration (370), geography (307), management science and engineering (114), applied economics (75), ethnology (58) and biology (55).

In order to further understand the background of the secondary disciplines involved in the doctoral study in tourism, secondary discipline backgrounds of doctoral dissertation researchers were compared among business administration, geography, applied economics, ethnology and biology (there is no secondary discipline under management science or engineering). As shown in Figure 8.6, tourism management (247) and human geography (235) are important secondary disciplines and professional platforms for tourism research. In addition, doctoral students with the background of corporate management, physical geography and ecology are also more involved in tourism research.

Figure 8.7 further analyses the variation trends of doctoral dissertations produced by the two secondary disciplines of tourism management and human geography and the first-level discipline of management science and engineering over the years. PhD students in human geography and management science and

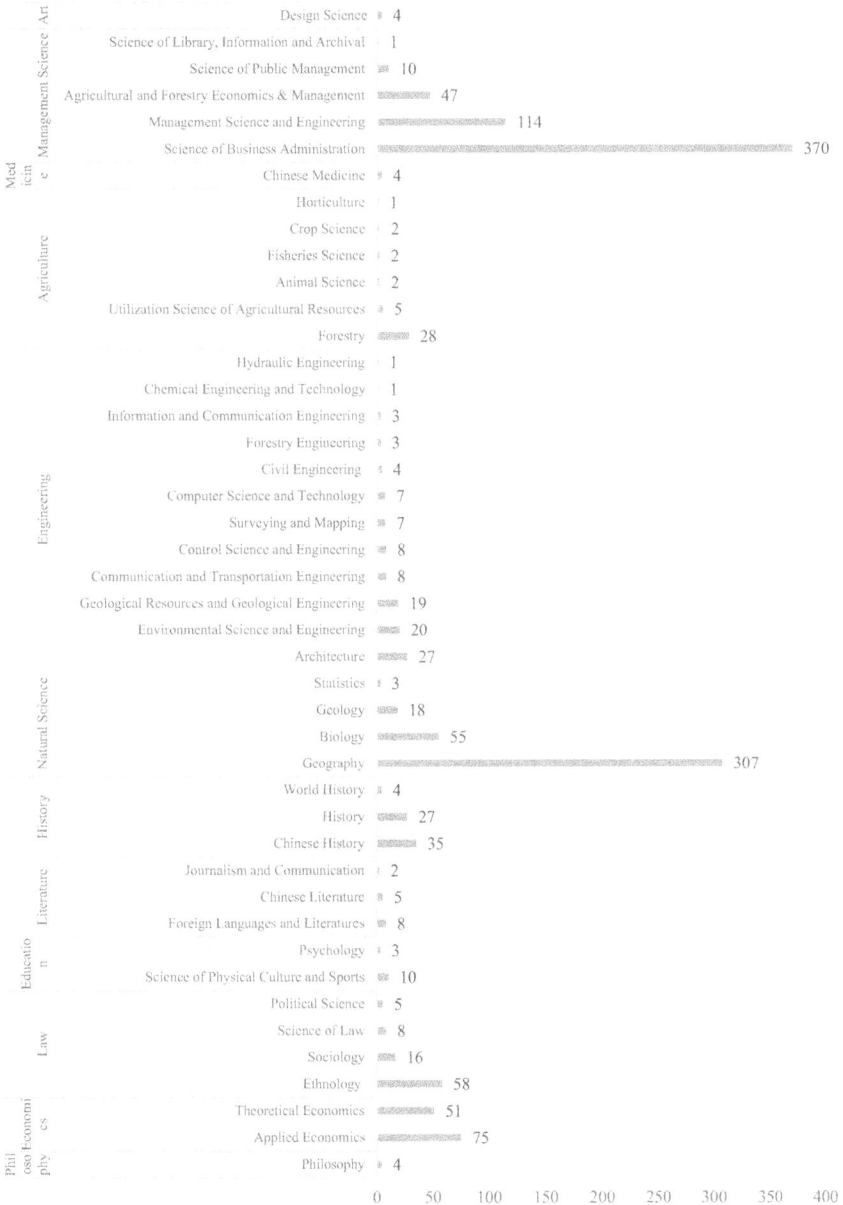

Discipline	Value
Design Science	4
Science of Library, Information and Archival	1
Science of Public Management	10
Agricultural and Forestry Economics & Management	47
Management Science and Engineering	114
Science of Business Administration	370
Chinese Medicine	4
Horticulture	1
Crop Science	2
Fisheries Science	2
Animal Science	2
Utilization Science of Agricultural Resources	5
Forestry	28
Hydraulic Engineering	1
Chemical Engineering and Technology	1
Information and Communication Engineering	3
Forestry Engineering	3
Civil Engineering	4
Computer Science and Technology	7
Surveying and Mapping	7
Control Science and Engineering	8
Communication and Transportation Engineering	8
Geological Resources and Geological Engineering	19
Environmental Science and Engineering	20
Architecture	27
Statistics	3
Geology	18
Biology	55
Geography	307
World History	4
History	27
Chinese History	35
Journalism and Communication	2
Chinese Literature	5
Foreign Languages and Literatures	8
Psychology	3
Science of Physical Culture and Sports	10
Political Science	5
Science of Law	8
Sociology	16
Ethnology	58
Theoretical Economics	51
Applied Economics	75
Philosophy	4

Figure 8.5 Comparison of discipline backgrounds of doctoral dissertation researchers (first-level disciplines).

Data source: The author collated the data from CDFD on the Wanfang Data Knowledge Service Platform.

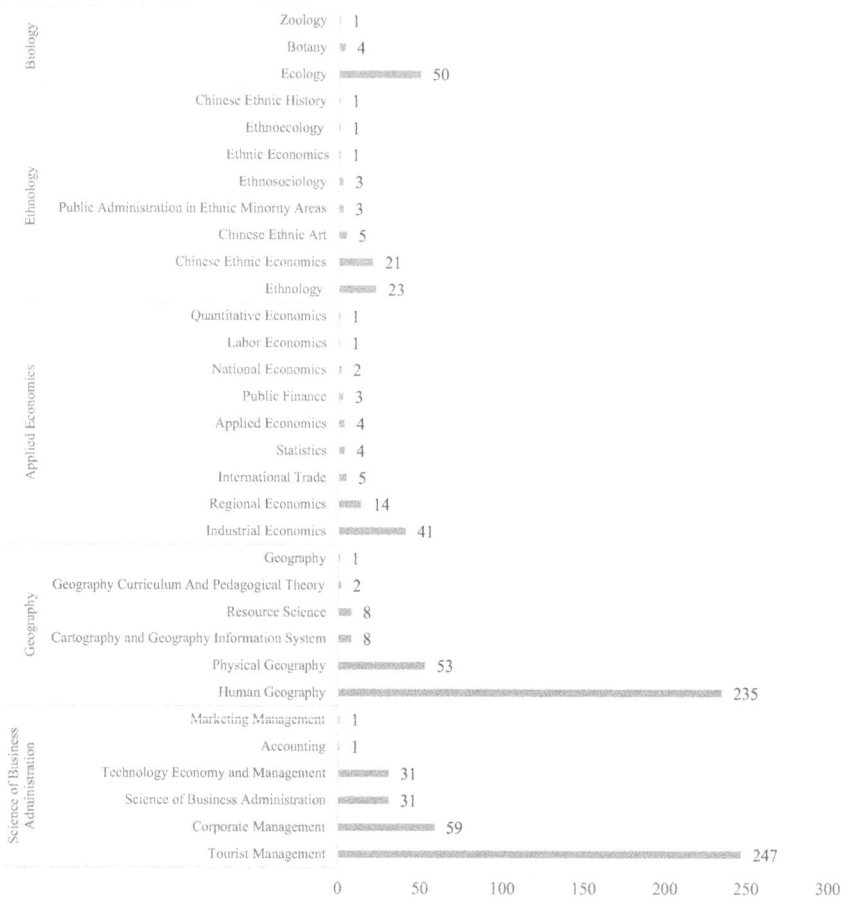

Figure 8.6 Comparison of discipline backgrounds of doctoral dissertation researchers (second-level disciplines).

Data source: The author collated the data from CDFD on the Wanfang Data Knowledge Service Platform.

engineering were engaged in tourism studies earlier. The first doctoral dissertations in tourism management appeared in 2002 after the establishment of independent tourism management doctoral programs. From 2006 to 2011, tourism management dissertations had a rapid growth. However, dissertations on tourism by PhD students in human geography appeared to decline slowly and those in management science and engineering fluctuated but remained relatively low.

Research topics of doctoral dissertations

The research hotspot is a scientific problem or topic jointly studied by a number of journals and literature with internal logical connection during a certain

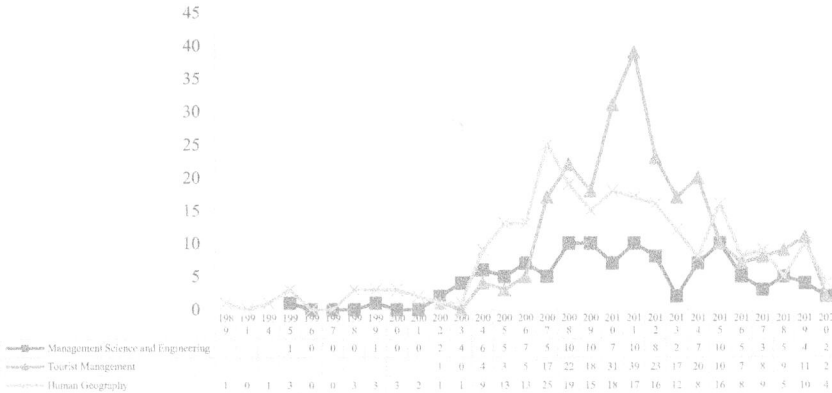

Figure 8.7 Trends of doctoral dissertations produced by the disciplines of tourism management, human geography, corporate management, physical geography and ecology over the years.

Data source: The author collated the data from CDFD on the Wanfang Data Knowledge Service Platform.

period of time (Chen 2004a). Keywords are the author's highly condensed and summarised content of the article, which can reflect the core theme of the article to a certain extent. Therefore, the analysis of the paradigmatic relations of keywords in literature in a certain field can reveal the academic research hotspots in this field (Chen 2004a). Based on the keywords of 1,395 doctoral dissertations (1989–2020) collected from the CDFD database, this paper uses CiteSpace 5.7.R5 software to perform visual analysis to determine the research hotspots of doctoral dissertations in tourism.

Due to some lack of keyword information in data, CiteSpace 5.7.R5 software eventually identified 1,382 dissertations from 1,399 dissertations for analysis. In addition, considering the long time span of the data, the time slicing was set at three years. The knowledge graph was plotted using the pathfinder algorithm. In order to more clearly show the characteristics of the research focus, this study combined and deleted the nodes with similar meanings. For example, nodes with little research significance (e.g. tourism and tourism industry) were deleted as well as nodes with information on case location (e.g. China, Thailand, Mount Tai). Subsequently, the keyword co-occurrence network was finally obtained with 381 nodes and 556 links. The node data displayed in the graph (Figure 8.8) was selected according to the threshold value set by DGREE = 5.

In the keyword co-occurrence network (Figure 8.8), each ring represents the evolution of a certain keyword. The colour of the ring represents the time of the emergence of the corresponding keyword and the size of the ring represents the frequency of the occurrence of the keyword. The larger the ring, the greater the frequency. Therefore, keywords with higher occurrence frequency can represent the focus of PhD-level tourism research to a certain extent. As

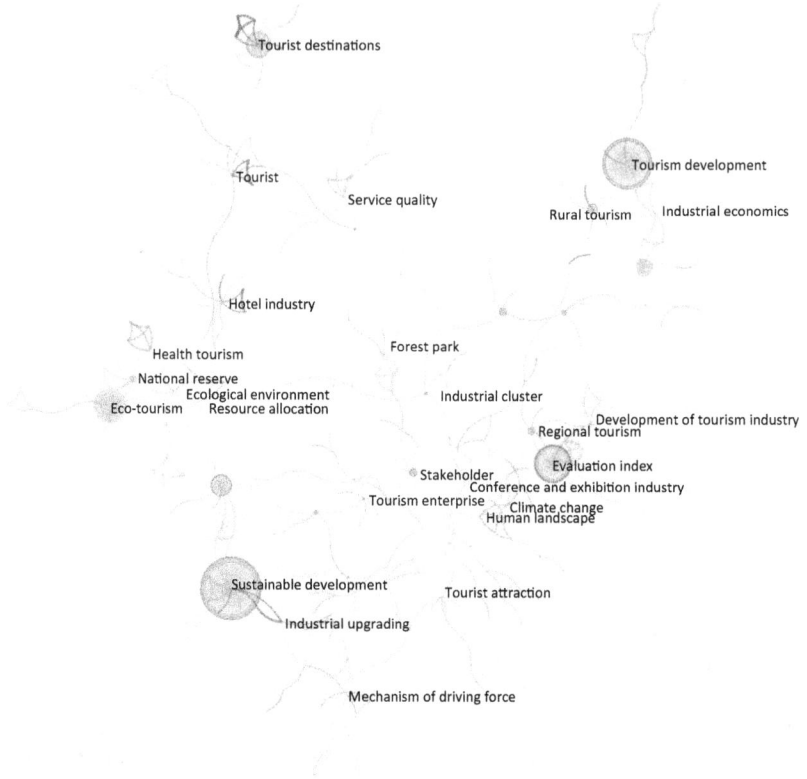

Figure 8.8 Keyword co-appearance network (n = 1,382).

shown in Figure 8.8, keywords with higher frequency are sustainable development, tourism development, evaluation index, ecotourism and tourism destination (also see Figure 8.9).

In this study, the list of the top ten keywords in order of centrality was also obtained by using the CiteSpace 5.7.R5 software (Table 8.2), so as to measure the hot research fields in tourism doctoral dissertations more scientifically and effectively. Centrality is mainly used to measure the effect intensity of nodes in a network. The higher the centrality, the stronger the importance and influence of the node in the whole network (the node whose centrality exceeds 0.1 is called the key node) and the more likely it is to establish a co-occurrence relationship with other nodes.

As seen in Table 8.2, the centrality of industrial cluster is 0.39, ranking first in the sequence and making it the core node of the entire research. It, together with resource allocation and tourism resources (with a centrality of 0.33 and 0.29, respectively), are the key nodes of this map. They are also an important knowledge base in the field of PhD-level tourism research and have the closest connection with other keywords.

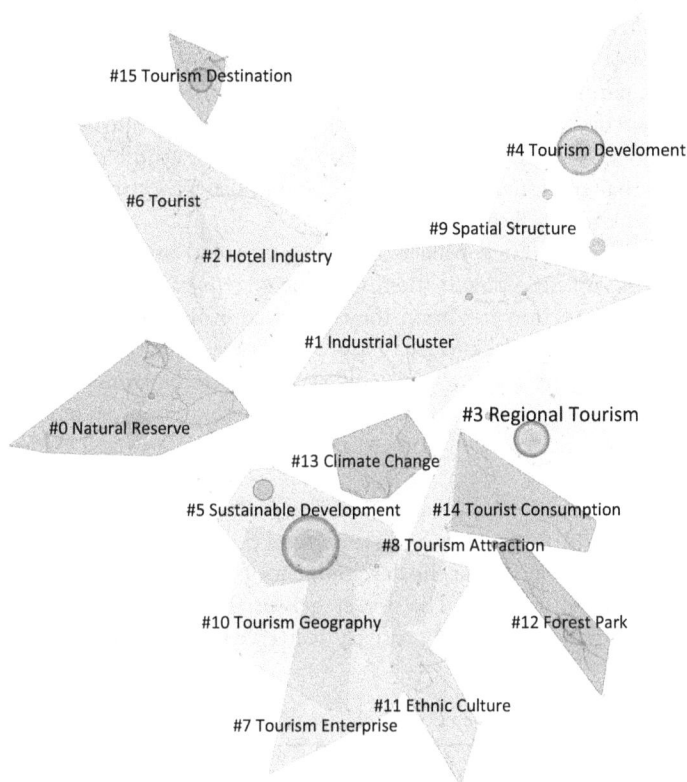

Figure 8.9 Keywords clustering knowledge graph (n = 1,382).

Table 8.2 List of the top ten keywords sorted by centrality

No	Keywords	Centrality	Frequency	Year
1	Industrial cluster	0.39	10	2007
2	Resource allocation	0.33	6	2005
3	Tourism resources	0.29	56	2000
4	Conference and exhibition industry	0.27	4	2009
5	Market positioning	0.26	2	2009
6	Economic development	0.23	3	2007
7	Evaluation index	0.22	71	2002
8	Industrial structure	0.21	10	2007
9	Regional tourism	0.18	19	1998
10	Strategic planning	0.18	9	2006

In order to conduct a better comprehensive exploration of keywords, this study adopts a log-likelihood ratio (LLR) to carry out clustering analysis on keywords, thus generating a keyword clustering knowledge graph as shown in Figure 8.10. Judging from the values of the graph, the Modularity Q is 0.8808 (>0.3 is significant construction) and the weighted mean silhouette S is 0.9545(>0.5 is reasonable). Both values are within a reasonable range, indicating that the clustering effect of this study is significant. These clusters reflect the development status and hotspot of PhD-level tourism research in China. Sixteen cluster labels were included: natural reserves, industrial cluster, restaurant industry, regional tourism, tourism development, sustainable tourism, tourists, tourism enterprise, tourism attraction, spatial structure, tourism geography, ethnic culture, forest park, climate change, tourist consumption and tourism destination. Specific clustering information is shown in Table 8.3.

In order to more accurately reflect the changing trend of the keywords, the time slice was set at one year and the rest of the system parameters were set to match the previous section to obtain the time zone evolution map of the keywords (Figure 8.10). The black font in Figure 8.10 represents the keywords (the keyword displayed in the graph was selected according to the threshold value set by DGREE = 5). The size of the circular node represents the frequency of the keywords and the line represents the co-occurrence relationship between the keywords.

The time zone evolution map can reflect the evolution path of hot topics in different time periods. By observing the time zone evolution map representation of keywords, it can be found that the development of PhD-level tourism research can be divided into four stages. The first stage was from 1989 to 1997, during which time the early research was relatively scattered and had not yet formed a

Table 8.3 Keywords co-occurrence network cluster table

Cluster ID	Size	Silhouette	Top three terms
0	28	0.961	Natural reserves; ecological tourism; health tourism
1	25	0.884	Industrial cluster; economic growth; tourism management
2	24	1	Restaurant industry; economic gain; lodging industry
3	23	1	Regional tourism; tourism development; development of tourism industry
4	23	1	Tourism development; tourism economy; resource development
5	22	0.946	Sustainable tourism; tourism resources; tourism demand
6	21	0.919	Tourists; consumption behaviour; community engagement
7	19	0.918	Tourism enterprise; city tour; evaluation index
8	16	0.891	Tourism attraction; strategic management; resort area
9	15	0.927	Spatial structure; rural tourism; tourist experience
10	15	0.972	Tourism geography; tourist area; tourism culture
11	14	0.989	Ethnic culture; dynamic mechanism; mode
12	14	1	Forest park; tourist satisfaction; stakeholder
13	13	0.966	Climate change; tourism product; human landscape
14	12	0.969	Tourist consumption; input-output; test the model
15	11	0.918	Tourism destination; inbound tourist; sightseeing place

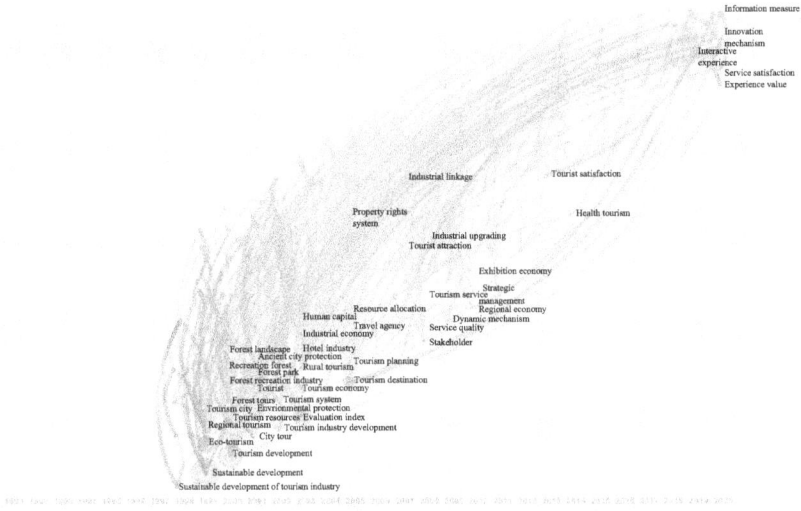

Figure 8.10 Keywords time zone distribution map (n = 1,382).

scale of high-frequency words. During the second stage, from 1998 to 2003, the number of keywords showed a rapid growth trend, and PhD-level tourism research entered an active period. Tourism development, especially sustainable development, was the research hotspots at this stage. The research focused more on the evaluation of tourism development and the design of the corresponding evaluation index system. Different types of tourism, such as ecotourism, forest tourism, city tour, rural tourism and regional tourism entered the field of vision of PhD-level tourism research. In addition, the research related to the hotel industry also began to attract the attention of PhD students, especially as it involved human capital.

In the third stage, from 2004 to 2011, the number of keywords studied was large, representing the peak of the annual number of keywords in the past two decades. The research on hotel industry and rural tourism had been further developed, focusing on hotel service quality and stakeholders, respectively. Topics related to the tourism industry (especially the tourism economy such as industrial linkage, industrial upgrading and industrial economic development), conferences and exhibitions as well as the development of tourist destinations and tourist attractions had become new research hotspots in PhD-level tourism research. The fourth stage spanned from 2012 to 2020. Although the time span of this stage is large, the number of keywords in PhD-level tourism research decreased rapidly as the popularity of related research topics gradually decreased. However, there are still some emerging topics in the field of doctoral tourism research, such as the health tourism and the tourism experience.

Outstanding issues

PhD research in tourism has been growing in both quantity and quality, creating some outstanding issues. As discussed above, PhD programs in China are managed in a hierarchical order and PhD programs in tourism management as a secondary-level discipline has not obtained sufficient resources and recognition in many universities. Some are shrinking due to lack of support, which is particularly true in research-oriented universities. Although efforts have been made in recent years by tourism academics to upgrade tourism management PhD programs to first-level disciplines, as of 2020, no approvals have been obtained from the Ministry of Education of the PRC. Thus, PhD programs in tourism management might find it necessary to contract in the future. However, the number of PhD students in tourism fields will not necessarily decrease because other first-level disciplines can engage in PhD studies on tourism.

Another issue is related to publication pressures for PhD students. Although journal publications are not required by the State Education Commission of the PRC, more and more universities require students to publish in prestigious Chinese journals before they can obtain a degree. There are two reasons behind this requirement. One is to increase the competitiveness of the graduates in the job market, while the other is to increase PhD students' contributions to their universities. In addition, since the total number of papers that can be published in prestigious Chinese journals is limited, PhD students are attempting to publish in international journals. As such, PhD candidates are under the pressure of two major tasks, journal publications and their dissertations, which do not necessarily support each other.

The debates over research problematisation are still ongoing. In the beginning, PhD dissertations were intended to address practical issues by identifying places and resources that had potential for tourism development (Bao, Huang & Chen 2019; Bao & Ma 2011). Thus, many studies were policy papers aiming to provide a solution to a practical issue, and it took quite a long time to educate young researchers to differentiate practical problems from research problems. The workshop series jointly held by Sun Yat-sen University and Hong Kong Polytechnic University has helped many young scholars and PhD students have a better understanding of the difference. Currently, debates continue as to whether research problems should be practice-oriented or theory-oriented and how PhD work can make a contribution to practical problems or theoretical problems through research (Holmström, Ketokivi & Hameri 2009; Xu, Ding & Packer 2008).

Gender equality is also an emerging issue. Tourism is a gendered landscape, and this is also reflected in tourist education (Pritchard & Morgan 2000) where females dominate the field (Munar et al. 2017). This dominance is observed in undergraduate, master's and doctoral education. However, once these female academics begin their academic careers at universities or research institutions, the increasing pressure in the academic world for research brings additional barriers for them to share an equal status with their male counterparts. Since research work conducted by Chinese academics often requires incursions into one's leisure time, and women are generally burdened with more family responsibilities, the time that women academics can devote to research, bid for research grants and publish is comparably less than their male counterparts (Liu, Wang & Xu 2019; Xu, Wang & Ye 2017). As a result, women generally do not perform as well as men in many research performance indicators, showing that they are in a weak position for career progression and promotion.

Conclusion

Tourism higher education has expanded significantly since the 1980s, which is aligned with tourism industry development and the expansion of higher education in China. Tourism PhD programs are also expanding. To date, doctoral tourism education has been offered within the disciplines of geography, management, economics and other first-level disciplines. The cultivation of tourism PhD students in China has formed on a large scale and entered a period of rapid development.

With the development of tourism management as an independent discipline, tourism PhD education in China has gradually stepped away from the original discipline platforms such as human geography and management and turned to tourism management as its own discipline. The independent PhD education of tourism management has also formed a doctoral training system with different characteristics. With the standardisation development of higher education in China, tourism PhD education has entered connotative development. Many comprehensive universities have paid more attention to teaching materials, teachers, teaching implementation and educational means to provide sufficient impetus for the construction and improvement of China's tourism higher education system.

The higher education system has been able to carry out relatively mature and high-level tourism research in the tourism management discipline. Due to the strict quality control of the doctoral dissertations in China's higher education system, especially the effect of the anonymous review and pre-defence system, the quality of tourism dissertations has significantly improved. Therefore, the doctoral dissertations in tourism research also make a great knowledge contribution to the field.

In general, the tourism research field at the PhD-level has a high degree of overall activity and involves a wide range of disciplines and professional levels, highlighting the interdisciplinary characteristics of the field, mainly including the disciplines of tourism management, corporate management, physical geography, ecology, economics and so on. The hotspots of tourism research at the PhD level focus on natural reserves, industrial clusters, the restaurant industry, regional tourism, sustainable tourism development, tourists, tourism enterprises, tourism attractions and tourism destinations. New hotspots have emerged over time, from traditional research topics, including the early evaluation index system on tourism development to different forms, from the corresponding research on the hotel and exhibition industries to the new multiple composite research topics, such as health tourism and interactive experiences. Theories of other disciplines are continually being integrated.

In recent years, although the quality of Chinese doctoral education has been improving and the research is being carried out rigorously and is recognised by the academic society and industry, there are still many problems in doctoral tourism education. Still, these challenges and issues can lead tourism educators to think in-depth about the future of doctoral tourism education, leading to higher-quality programs.

Notes

1 In some places which are generally municipalities directly under the central government, due to institutional settings and local special circumstances, the education administrative departments are called *education commissions*, for example, Beijing Municipal Education Commission, Tianjin Municipal Education Commission, Shanghai Municipal Education Commission and Chongqing Municipal Education Commission. Their functions are equivalent to those of other provincial education departments.
2 In 1998, the State Education Commission of the PRC was renamed the Ministry of Education of the PRC.

References

Bao, J 2002, 'Tourism geography as the subject of doctoral dissertations in China, 1989–2000', *Tourism Geographies: An International Journal of Tourism Space, Place and Environment*, vol. 4, no. 2, pp. 148–152. Available from: Taylor & Francis Online. [28 August 2020].

Bao, J, Huang, S & Chen, G 2019, 'Forty years of China tourism research: reflections and prospects', *Journal of China Tourism Research*, vol. 15, no. 3, pp. 283–294. Available from: Taylor & Francis Online. [28 August 2020].

Bao, J & Ma, LJ 2011, 'Tourism geography in China, 1978–2008: whence, what and whither?', *Progress in Human Geography*, vol. 35, no. 1, pp. 3–20. Available from: SAGE Journals. [28 August 2020].

Chen, C 2004a, 'Searching for intellectual turning points: progressive knowledge domain visualization'. *Proceedings of the National Academy of Sciences*, vol. 101, no. suppl 1, pp. 5303–5310. [28 August 2020].

Chen, D 2004b, 'The development of tourism science from the perspective of doctoral dissertation of tourism studies', *Tourism Tribune*, vol. 19, no. 6, pp. 9–14. Available from: https://kns-cnki-net. [28 August 2020].

Holmström, J, Ketokivi, M & Hameri, AP 2009, 'Bridging practice and theory: a design science approach', *Decision Sciences*, vol. 40, no. 1, pp. 65–87. Available from: Wiley Online Library. [28 August 2020].

Huang, S 2011, 'Tourism as the subject of China's doctoral dissertations', *Annals of Tourism Research*, vol. 38, no. 1, pp. 316–319. Available from: CAB Direct. [28 August 2020].

Liu, F, Wang, H & Xu, H 2019, 'Gender differences in academic output of Chinese tourism scholars: based on literatures published in Tourism Tribune', *Tourism Tribune*, vol. 34, no. 12, pp. 109–119. Available from: https://kns-cnki-net. [28 August 2020].

Meyer-Arendt, KJ 2000, 'Commentary: tourism geography as the subject of North American doctoral dissertations and master's theses, 1951–1998', *Tourism Geographies*, vol. 2, no. 2, pp. 140–156. Available from: Taylor & Francis Online. [28 August 2020].

Meyer-Arendt, KJ & Justice, C 2002, 'Tourism as the subject of North American', *Annals of Tourism Research*, vol. 29, no. 4, pp. 1171–1174. Available from: http://dx.doi.org/10.1016/S0160-7383(02)00038-5. [28 August 2020].

Munar, AM, Khoo-Lattimore, C, Chambers, D & Biran, A 2017, 'The academia we have and the one we want: on the centrality of gender equality', *Anatolia*, vol. 28, no. 4, pp. 582–591. Available from: Taylor & Francis Online. [28 August 2020].

Pritchard, A & Morgan, NJ 2000, 'Privileging the male gaze: gendered tourism landscapes', *Annals of Tourism Research*, vol. 27, no. 4, pp. 884–905. Available from: ScienceDirect. [28 August 2020].

UNESCO Institute for Statistics 2012, *International Standard Classification of Education: ISCED 2011*, Montreal, UNESCO Institute for Statistics. Available from: http://uis.unesco.org/en/topic/international-standard-classification-education-isced. [28 August 2020].

Weiler, B, Moyle, B & McLennan, CL 2012, 'Disciplines that influence tourism doctoral research: the United States, Canada, Australia and New Zealand', *Annals of Tourism Research*, vol. 39, no. 3, pp. 1425–1445. Available from: ScienceDirect. [28 August 2020].

Xu, H, Ding, P & Packer, J 2008, 'Tourism research in China: understanding the unique cultural contexts and complexities', *Current Issues in Tourism*, vol. 11, no. 6, pp. 473–491. Available from: Taylor & Francis Online. [28 August 2020].

Xu, H, Wang, K & Ye, T 2017, 'Women's awareness of gender issues in Chinese tourism academia', *Anatolia*, vol. 28, no. 4, pp. 553–566. Available from: Taylor & Francis Online. [28 August 2020].

Ying, T & Xiao, H 2012, 'Knowledge linkage: a social network analysis of tourism dissertation subjects', *Journal of Hospitality & Tourism Research*, vol. 36, no. 4, pp. 450–477. Available from: SAGE Journals. [28 August 2020].

9 Curriculum settings and comparisons

Chaozhi Zhang and Xiaofeng Zhou

Introduction

With a rising disposable income and rapid development of infrastructure such as a high-speed rail network (despite the prevailing Covid-19 situation), China is expected to become one of the largest domestic and international tourism markets in the coming years. Concurrently, China is the biggest tourism education market in the world (Bao & Zhu 2008; Zhang 2015). However, China's tourism education still has a significant gap in terms of international competitiveness when compared to other countries. To identify the existing gaps in China's tourism education system, it is conducive to compare the curriculum system in the various Chinese tourism colleges against the current systems in the Western countries and other parts of the world.

Chinese tourism education had its beginnings in the 1980s. After nearly 40 years of development, Chinese tourism education became systematic and scaled up to meet demand due to the rapid growth of China's tourism and education industry. By the end of 2017, there was 2,641 higher education institutions offering tourism programs (including tourism colleges and colleges with tourism-related majors), with 608 universities offering undergraduate degrees, accounting for 23% of the total. A total of 274,000 students were enrolled in tourism management majors nationwide (Tong & Lu 2019), and a large number of tourism professionals were trained to support the development of the tourism industry. The tourism industry's "Twelfth Five-Year" Talents Plan (2011–2015) established "the strategic position of priority for tourism talents in the development of tourism industry" and the strategy of "increasing the talent pool for tourism development and striving to establish competitive tourism professionals, thereby cultivating a large-scale, high-quality and reasonably structured talented professionals group to match the development of tourism" (Ministry of Culture and Tourism of the People's Republic of China, 2011).

However, there is always controversy among comprehensive and research universities on setting up a major practical curriculum such as tourism management, including research-oriented courses and students training. Many issues such as "program adjustment," "theoretic insufficiency," "shortage of teaching materials," and "contradictory major positioning" exist (Su & Yin 2005). There are

DOI: 10.4324/9781003004363-9

also situations where the training objectives do not match the market needs, resulting in unsatisfactory training outcomes. Therefore, this study is a practical guide targeted at comparing the training goals, setting concepts, and course contents between well-known Chinese and foreign tourism colleges and universities, and providing an ideological basis for the curriculum setting for Chinese tourism colleges as well as a reference for the positioning of professional training.

Literature review

Core idea of Chinese tourism management major curriculum

Scholars have suggested the following basic guidelines for the domestic tourism management major curriculum setting.

The curriculum setting system should aim at training interdisciplinary talents

The developmental history of international and domestic tourism showed that tourism development mainly requires two types of professionals: (1) professionals for development, including regional tourism development, tourist attractions development, tourism product development, and tourism market development; and (2) professionals for management, including tourism industry management and tourism business management, tourist attractions management, and tourism facility management (Xu & Zhang 2004). A three-series curriculum system comprising core curriculum, skills course, and related courses is more common in Chinese tourism colleges (Luo & Luo 1997), focusing on training interdisciplinary talents (Yue & Li 2009). Based on the two major groups of compulsory courses and electives, the curriculum is divided into "public," "professional basic courses," "professional compulsory courses," "electives," and "public electives" (Guo & Yang 2008) under the framework of general courses (or broad core curriculum, which consists of standard courses + introductory courses) – professional courses – additional courses (i.e., expanded courses or related courses) (Lu 1999). A variety of course types such as "module" have also been proposed and implemented, including "core course modules," "professional course modules," and "expanded course modules" (Zhao & Wang 1998).

Management courses account for a large proportion

According to the discipline's guidance from the Education Department, the undergraduate tourism management courses are under the first level of business management. Based on this, four "series types" of the curriculum are designed. There are the standard core course series, which include the "general management" courses, the "tourism management core" course series, the "tourism management professional" course series, and the "modern management talents" educational series (Lin 1998). Based on the statistical analysis of the curriculum documents of 69 colleges offering tourism majors in China, the management courses top the table, emphasizing the theoretical basis of tourism (Sun 2013).

Integration of tourism teaching and scientific research

Tourism management research results have been applied to the tourism curriculum in Chinese colleges (Yuan et al. 2005). By targeting the existing issues in the lower level of the curriculum, several points have become pertinent for curriculum setting (Xu 1999). These include connecting curriculum content and industry needs, the integrity of the curriculum system, the connection between tourism education and economic development, and the uniqueness of the training. A curriculum system should emphasize teaching, practice, and research including theoretical basis, applied courses, and practicums (Zeng & Peng 2008). The tourism management discipline should reinforce the summary of the results from tourism development projects and research institutes, thereby taking advantage of research outcomes (Wang, Hong & Niu 2016) to provide the basis for the curriculum setting.

Main problems in the tourism curriculum setting

At present, the following problems exist in the curriculum setting of China's tourism management major.

Course structure and professionalism

At present, the contents of Chinese courses are unclear, textbooks are outdated and unpractical, and out of line with industrial requirements, with these identified failings being the apparent reasons for lagging behind courses in foreign institutions (Zhang 1998; Jiang, Wu & Huang 2003). The curriculum emphasizes unity, with a characteristic of content repetition, and lacks content variety. Due to the differences in professional teaching, the courses are set based on subjects and skills sets of teachers, while the core courses are inconsistent with no clear teaching outcomes (Liu 2003). The major curriculum model settings tend to be outdated and lack innovation (Qian 2005). As a result, the professional curriculum content setting is limited to a relatively narrow field, lacking relevancy to the tourism industry resulting in an inadequate curriculum system (Tong & Lu 2019).

Training objective and model

Earlier studies have shown that teachers are always dominant in teaching and imparting knowledge in their subject areas. However, students' potential may not be fully developed, given the narrow range of knowledge learned and overdependency on teachers (Bo & Yao 1999). Students are short on practical work-integrated learning experiences (practicums) due to the "blind" implementation of modern educational tools and outdated teaching methods, resulting in graduates failing to meet the employment requirements of tourism management positions (Liu 2004). As a result, the poor positioning of the tourism management curriculum system with obsolete teaching models causes a disjoint between the course settings and training goals (Feng & Li 2007). With the recent transformation of the Chinese tourism industry from being extensive and inefficient to organized and efficient, tourism management teaching has gradually evolved in alignment with the initial intention of training goals. However, it still does not match the development of the tourism industry (Wang, Hong & Niu 2016). This incongruency creates practical difficulty to meet the training demands of tourism professionals ranging from issues in types and number of courses, the proportion of teaching categories, study hours, and the integration between different courses. Therefore, the goals and contents of each professional course need to be further adjusted (Tong & Lu 2019).

Practicums in consideration of social needs

Graduates lack independent learning abilities and problem-solving skills due to poor connections between the Chinese tourism management curriculum setting and practical experience (Zhuang 1998). Tourism education in China emphasizes theory more than applied learning with the added difficulty of securing internships. This issue attracted the attention of many experts in the academic field. On-campus practicum is often hard to implement; however, an off-campus internship is also challenging to secure because the university is poorly connected to the industry and has difficulties securing internships for students. These operational issues are the fundamental obstacles to students' lack of applied skills (Wang 2000; Tan & Mao 2001; Li & Ji 2007). The internship and skills courses arrangement is very vague; the "practical teaching" courses are often inadequate while the "skills-oriented" courses cannot create opportunities for students to learn "hands-on skills" (Guo, Ma & Liu 2007). The situation is improving as specific colleges are receiving some form of support from industry partners. However, these industry collaborators usually provide certain guarantees in terms of funding support. They do not provide in-field placements which do not solve the main problem of a shortage in practicums or internships (Gong & Chen 2018).

Knowledge system, theoretical paradigm, and research methods

Tourism education in China face problems such as monotonous theoretical paradigm, unsystematic research methods, unstandardized keywords in the tourism research system, and the imperfect academic criterion (Xie 2003). It is challenging to integrate the tourism knowledge system's constructional requirements into teaching and form a unified understanding of the tourism educational knowledge system and impart such knowledge to students (Li & Zhang 2010). At the same time, these courses tend to have repetitive content and lack distinct teaching outcomes. In light of this situation, the summary and refinement of the syllabus cannot predict and analyze market demand and market outlook.

After a decade of developing and fine-tuning the tourism management programs' curriculum, some of the problems mentioned above have been addressed. For example, in terms of the quality of teaching staff, the previously mentioned problems such as lack of practical experience, narrow expertise, and old-fashioned teaching methods have been primarily addressed. Nevertheless, there are still problems that cannot be ignored in the curriculum setting. First, there is a need for confluence between the scientific curriculum settings and training goals alongside unique professional characteristics and correct market positioning; second, skill training needs to be further improved.

Research methods

Sample selection

In order to make the study sample more comparable, the top 18 Chinese tourism management colleges, the top 20 overseas tourism colleges, and well-known colleges of tourism management majors worldwide are selected as typical samples to identify the curriculum setting and overall characteristics. Furthermore, a comparative analysis is done to guide Chinese colleges to optimize the curriculum scientifically and systematically.

China's college tourism curriculum can be divided into two parts: academic content and practice. Based on the commonly used curriculum classification methods in Chinese teaching management (Xu, Xu & Wang 2016), the curriculum is divided into "public courses," "specialized basic courses," and "specialized core courses."

Public courses include: Situation and Policy, Ideological & Moral Cultivation and Fundamentals of Law, English, Computer, Military Theory and Practice, Employment Guidance, and so on;

Specialized introductory courses include: Basic Tourism Knowledge, Economics of Tourism, Human Geography, Introduction to Tourism, Sociological Research Methods, and Advanced Mathematics;

Specialized core courses include the main component of the tourism management curriculum system, including specialized compulsory courses. This part introduces the relevant theories of tourism disciplines systematically and thoroughly, requiring students' mastery. Specific courses include Tourism Management, Hotel Management, Tourism Marketing, Tourism Scenic Landscape Management, and so on. There are also flexible specialized elective courses, such as Tourism Folklore, Tourist Attraction Design, Tourism Route Design, and Tourism Resources and Development, which are provided to cater to students' interests and faculty.

Applied skills are an essential part of the curriculum with components such as practicum and internship with the objective of enhancing the student's professional abilities and qualities through training of specific skill sets.

Based on the ranking of China's tourism management majors in 2020 taken from "Network of Science & Education Evaluation in China" website (Gerenjianli 2019) and "the Ranking of Academic Subjects 2020 – Hospitality & Tourism Management" from Shanghai Ranking's Global Ranking of Academic Subjects (Shanghai Ranking Consultancy 2020), this study selected the top 6 from the top 20 Chinese universities on the subject of Tourism Management for comparative analysis. They are Sun Yat-sen University, Fudan University, Beijing International Studies University, Dongbei University of Finance and Economics, East China Normal University, and Hubei University (see Appendix 1).

According to the 2019 QS World University (Quacquarelli Symonds 2019) ranking in "Hospital and Leisure Management" (updated on March 14, 2019) and the 2020 Shanghai Ranking's Global Ranking of Academic Subjects (Shanghai Ranking Consultancy 2020), this study selected 14 overseas universities for comparative analysis from the top 20 universities in the subject area of tourism and leisure management. These institutions are Hong Kong Polytechnic University, Griffith University, University of Surrey, Pennsylvania State University-University Park, University of Queensland, Bournemouth University, Temple University, Washington State University, Kyung Hee University, Florida State University, University of Nevada – Las Vegas, University of South Carolina – Columbia, the University of Florida, and the University of Houston.

Characteristics of Chinese tourism institutions' curriculum setting

The curriculum is mainly influenced by training objectives both holistically and specifically. The curriculum of Chinese tourism colleges is shown in Tables 9.1 and 9.2. By comparing and analyzing the introductory course content and knowledge structure, conclusions are drawn as follows.

Course contents

Course contents reflect academic backgrounds and highlight the core development of the industry

The foundation of Chinese tourism college curriculum contains diverse introductory courses: Management, Foreign Languages, Economics, Human Geography, and so on, which reflect not only the interdisciplinary characteristics of the tourism major but also highlight the direction of different course content settings. Most colleges have set up tourism management under the umbrella of first-level business management discipline in curriculum development. The core courses mainly focus on Management Science, combined with Tourism and Economics and integrated with related fields. It reinforces the basic building blocks of management and highlights the current characteristics of China's tourism industry development.

Table 9.1 List of curriculum structure of Chinese tourism colleges (number of courses)

No.	College Name	Professional Core Courses			Professional Specialized Courses				Public Courses		Practicum/ Internship	Total
		Tourism	Economics	Management	General Education	Tourism	Economics	Management	Humanities and Social Sciences	Natural Sciences		
1	Sun Yat-sen University	5	5	15	10	7	3	6	8	2	5	66
2	Beijing Second Foreign Language Institute	5	6	8	6	7	0	1	6	1	3	43
3	Dongbei University of Finance and Economics	6	6	5	9	11	7	8	18	6	8	84
4	East China Normal University	5	5	5	5	9	1	7	5	4	6	52
5	Hubei University	5	5	7	2	8	0	6	6	4	7	50
6	Fudan University	7	9	7	7	11	3	8	7	2	6	67
	Total	33	36	47	39	53	14	36	50	19	35	362
	Proportion	32.0%			39.2%				19.0%		9.7%	100%

Table 9.2 List of curriculum structure of Chinese tourism colleges (credits)

No.	College Name	Professional Core Courses			Professional Specialized Courses				Public Courses		Practical Internship	Total
		Tourism	Economics	Management	General Education	Tourism	Economics	Management	Humanities and Social Sciences	Natural Sciences		
1	Sun Yat-sen University	12	14	30	20	14	6	12	28	8	14	158
2	Beijing Second Foreign Language Institute	10	13	16	18	14	0	2	36	4	3	116
3	Dongbei University of Finance and Economics	13	20	12	19	22	16	16	35	21	22	196
4	East China Normal University	11	11	12	10	16	2	14	28	15	17	136
5	Hubei University	13	15	17.5	9	13.5	0	11.5	27	19	13	138.5
6	Fudan University	18	15.5	28.5	14	22	6	16	25	12	6	163
Total		77	88.5	116	90	101.5	30	71.5	179	79	75	907.5
Proportion		31.0%			32.3%				28.4%		8.2%	100%

Management courses such as Principles of Management Science, Service Management, Financial Management, and other management courses establish the central position of the curriculum and serve as the foundation for the specialized courses. Specialized core courses such as Introduction to Tourism, Tourism Geography, Tourism Psychology, and Tourism Resources and Development focus on the interdisciplinary characteristics of tourism and build a holistic and systematic knowledge network. Specialized electives such as Heritage Management and Tourism, Folklore and Tourism Development, Scenic Spots and Theme Park Management integrate different industrial perspectives and expand the scope of knowledge of related industries. These three parts work in unison to reflect the macro trends of globalization, commercial and cultural coordination, encompass the large body of knowledge involved in the tourism management discipline, represent all aspects of the tourism phenomenon, and embody the significant features of the tourism educational knowledge system.

The content of public courses is uniform and comprehensive

As the basis of specialized courses, the public compulsory courses impart generic knowledge to students in order to promote moral development, comprehensive skills, and knowledge in the disciplines. First, these courses help students cultivate moral and ethical awareness, fostering sustainable personal development capabilities and ground them for personal growth. Second, Tourism Management is an essential skill set in this fast-growing and globalized informational age. The inclusion of English within tourism majors further deepens the learning and practice runway of foreign languages. Additionally, Management, Economics, and industry rules pave the way for basic skills. Public courses such as those in natural sciences, humanities, and skills combined with specialized courses would lay the foundation for the students to refine their capability and growth.

Diversified internships accelerate the transformation of professional knowledge

As an applied major, the practice courses in Chinese colleges could be segregated into on-campus practice and off-campus internships. Internship refers to the practical application and enhancement of theoretical knowledge through practice. China's practice courses are typically designed with features of exposure, problem-based learning, and applied learning. Exposure refers to broadening students' minds by providing them with a "site" to practice their skills and apply their studied theories and, from there, be open to the industry and engage in building expertise. Problem-based learning refers to exposing students to solve problems in teams through cooperating, researching, "learning on the task," and communicating with others under the guidance of a teacher. Applied learning refers to putting creativity into practice, leading students to use and apply theoretical knowledge and developing their initial working ability and professional judgment.

Knowledge system

Focusing on management and building an interdisciplinary knowledge system

As a business management supplementary discipline, tourism management has its historical roots from the vocational training of hotel management. Its curriculum structure focuses on fundamental management knowledge and tourism-specialized knowledge. With the methods of theoretical study, practice, and basic research, a systematic framework of guided learning and autonomous learning can be built to cater to the new trend of tourism industry development. The knowledge system of tourism management includes the following features: (1) capitalizes on the advantages of its institutions to create a personalized and specialized knowledge structure under the overall development trend of tourism education; (2) multifaceted exchanges and collaboration between domestic and international institutions and improving professional training programs and curriculum. An open and cooperative demand-oriented curriculum setting combining production, education, and research has gradually been developed. Third, with basic knowledge of natural science, engineering, economics, psychology, and law, students are able to construct their own theories and methods for tourism resource development and management.

Capacity cultivation

Equal emphasis on both science and humanities incorporating theoretical knowledge and applied skills

The curriculum in Chinese tourism colleges fulfills China's needs in terms of national economic development in the 21st century. There is a priority in training students to understand and practice socialist core values to equip them with a sense of social responsibility and balanced mental health, including critical and scientific thinking and human accomplishment. The curriculum assists students to master the basic theory, knowledge, and skills of tourism management, gain familiarity with guidelines, policies, regulations, and developmental trends of China's tourism industry management. Furthermore, they help students acquire systematic expertise of management, great analysis ability and problem-solving skills, and innovation and entrepreneurship.

The current curriculum setting of tourism colleges has emphasis at different levels.

First, public introductory courses focus on cultivating students' personal qualities such as patriotism, scientific/rational worldview, and correct value system. These courses also require a high level of English, application skills, and basic knowledge of related laws to inspire students to manage modernization with a sense of responsibility, morality, and professional ethics.

Second, introductory and core specialized courses require students to master relevant knowledge and modern management methods, tourism, economics, and law. These courses prepare students by equipping them with management competencies, theoretical and practical abilities to adapt to the diverse needs of the tourism industry so that they are familiar with the various types of tourism subfields (tourism management, hotel management, exhibition or general management jobs) along with the capability for further study in scientific research institutions.

Third, practicums, academic visits, simulation, and other courses endow students with international perspectives, innovation, teamwork, leadership, communication, and applied skills such as skills to design tourism management system. In the next five to ten years of development, students will be able to engage in the development and management of tourism resources in medium- and large-sized companies strategically and innovatively.

Curriculum setting of internationally acclaimed tourism colleges

Since the 1980s, tourism has proliferated. Also, higher education has developed fast around the world (Strohberhn 1994; KohK 1995). Colleges in Europe have begun to create tourism courses which were introduced to higher education institutes. In Japan, vocational high schools have also started tourism-related courses and created a complete secondary to the undergraduate educational system (Xu & Zhang 2004). Nowadays, most top-rated tourism institutes are in Europe and the US. They represent the highest level of tourism management education till date. The following findings are identified concerning foreign tourism colleges' curriculum design (Tables 9.3 and 9.4).

Course contents

Emphasizing the international perspective and consolidating advantages status

International colleges typically focus on four subject areas: Business Management, Geography and Environmental Science, Culture and Human Science, and Art (Sun & Li 2019). These areas are very typical representations of modern education. The curriculum setting relies on the open economic situation and diverse culture of the country and region. International teaching staff brings more depth and perspectives. Based on the management and operation details of hotels, tourism, exhibitions, and so on, course settings typically focus on international hotel and tourism supply, demand, social system, and impact. The core objects of the industry are on the management-related courses (Wu, Tang & Cai 2002).

Moreover, setting up majors in "Hotels and Tourism Introduction" and "Hotels and Travel Management Organization" can give students a greater understanding of course contents and global industrial standards. Sustainability, environmental studies, tourism economics, and exhibition management

Table 9.3 List of the curriculum structure of the top 20 tourism institutions in the world (number of courses)

No	College Name	Professional Core Courses			Professional Specialized Courses				Public Courses		Practical	Total
		Tourism	Economics	Management	General Education	Tourism	Economics	Management	Humanities and Social Sciences	Natural Sciences	Internship	
1	The Hong Kong Polytechnic University	5	3	5	2	4	0	2	5	1	5	32
2	Griffith University	0	4	8	5	5	5	6	0	0	4	37
3	University of Surrey	5	4	4	6	2	4	4	0	0	1	30
4	Pennsylvania State University Park Campus	2	0	6	0	4	0	1	4	2	3	22
5	The University of Queensland	7	5	8	0	2	0	0	4	0	0	26
6	Bournemouth University	6	3	1	0	10	0	6	5	0	1	32
7	Temple University	8	5	3	0	6	0	7	6	2	3	40
8	Washington State University	3	3	14	5	0	0	0	0	0	10	35
9	Kyung Hee University	5	5	6	0	6	0	4	10	0	8	44
10	Florida State University	2	8	3	12	0	3	6	0	1	4	39
11	University of Nevada	0	4	4	7	0	0	0	11	1	6	33
12	University of South Carolina at Columbia	5	7	9	9	11	2	17	14	3	2	79
13	University of Florida	0	2	2	4	0	6	7	0	0	0	21
14	University of Houston	0	13	3	3	0	5	5	9	1	1	40
	Total	48	66	76	53	50	25	65	68	11	48	510
	Proportion	37.25%			37.84%				15.49%		9.41%	100%

Table 9.4 List of the curriculum structure of the top 20 overseas tourism institutions (credits)

No	College Name	Professional Core Courses			Professional Specialized Courses				Public Courses		Practical	Total
		Tourism	Economics	Management	General Education	Tourism	Economics	Management	Humanities and Social Sciences	Natural Sciences	Internship	
1	The Hong Kong Polytechnic University	14	9	15	0	12	0	6	18	3	18	95
2	Griffith University	0	40	80	50	50	50	60	0	0	40	370
3	University of Surrey	60	75	60	90	30	60	60	0	0	15	450
4	Pennsylvania State University Park Campus	6	0	23	0	12	0	3	12	6	7	69
5	The University of Queensland	14	14	16	0	4	0	0	8	0	0	56
6	Bournemouth University	120	60	20	0	200	0	120	80	0	20	620
7	Temple University	24	13	12	0	18	0	21	19	7	18	132
8	Washington State University	9	9	42	15	0	0	0	0	0	19	94
9	Kyung Hee University	15	18	15	0	18	0	12	27	0	27	132
10	Florida State University	6	24	9	36	0	9	18	3	0	28	133
11	University of Nevada	67							41	3	28	139
12	University of South Carolina at Columbia	75			27	33	6	21	49	18	25	254
13	University of Florida	0	6	6	12	0	18	21	0	0	0	63
14	University of Houston	0	39	9	10	0	16	17	33	12	25	161
	Total	2159							290	49	270	2768
	Proportion	78.00%							12.25%		9.75%	100%

demonstrate the interdisciplinary nature of tourism management programs. By introducing "Multi-Cultural Catering", "Airline Management," and "Attractions and Visitor Management" as elective courses, students can strengthen their understanding of career trajectories and the application of comprehensive knowledge and skills. As a result, by applying modern tourism management, hotel and travel activities management as course contents, graduates from these courses can contribute to the development of the tourism service industry.

A balance between natural and social sciences subjects

There is a boom in new marketing practices, resulting in high demand for small-scale businesses and specialized technologies. Public courses in natural science and social science, community development, and self-development create strong networks for students to expand their overall competencies. Such courses include the following areas: humanities, relationships, communities, and organizations under the globalization perspective. It also demonstrates the popularization of courses such as social and natural sciences, environmental study, applied computer technology, parks, and recreation that could intertwine with history and cultural studies and help students establish their worldviews. Developing leadership skills and personal growth, service-learning, and other courses could improve students' self-worth and aspirations. Public courses are often related to current events. By combining with professional requirement courses, students can understand the links between their courses and the tourism industry.

Practical internships highlight pragmatism; focus on individual vision and initiative

Relying on strong resource support, some international colleges have established internship arrangements with in-house hotels or industry partners to provide students with top-level, direct, and unique internship opportunities. The three benefits of internship include: (1) establishing cooperation with top hotels and tourist destinations around the world to provide students with a platform for applying theoretical knowledge and gaining applied skills and knowledge in line with industry development; (2) introducing a guidance system for specialized industrial personnel and build an internship curriculum that can lead students to pay attention to new theories and technologies within the hotel industry and tourism; (3) providing a broad range of career choices, combining personal interest and professional capabilities, and promoting experience accumulation and vision improvement (Hong Kong Programmes n.d.).

Knowledge system

The multidimensional integration: design curriculum flexibility

Foreign tourism colleges take into consideration problem-solving and challenges from the tourism operation, development, and management perspective. They include courses in basic skills, culture and humanities, professional knowledge and science, international perspectives, career recognition, and professional ethics under multiple dimensions. This integration forms a holistic knowledge structure based on the flexible setting of basic management modules, professional modules, and practical modules (Han & Lu 2010), and is divided into multiple disciplinaries, including but not exclusive to "Advanced Business," "Business and Human Resource Management," and "Business and Marketing." As a result, students could participate in the theory creation as well as application of tourism planning and development and concurrently explore conceptual and pragmatic issues in the hotel and tourism fields.

In terms of content formulation, tourism phenomena and trends are taken as the core body of knowledge building an overall structural framework around the knowledge content of tourism from the perspectives of globalization, cross-cultural diversity, and social, economic, political, geographic, environmental, and other fields (Huang 2009). From a hierarchical structure, the knowledge system highlights the practicability and the typical characteristics of industry development. It also avoids the repetition of theoretical intersections and forms problem-solving methods and approaches. Teaching methods focus on autonomous guidance, self-learning, and lifelong learning skills.

Capacity cultivation

To ensure effective training, relevant theories and applications of tourism planning and development needed to be covered in the courses. Against the background of global economic development, the curriculum content of foreign tourism colleges focuses on both local and international development trends, integrating practical content and methods with business and management as the core foundation, with emphasis on widening students' professional abilities and mindset:

First, the public courses connect students with the social environment, providing students the opportunity to recognize their personal value in the global tourism system and enhance their literacy and ability (School of Hotel and Tourism Management n.d.). Public courses also encourage students' learning motivation, cultivate creativity, responsibility, and long-term vision.

Second, the specialized introductory and core courses, including "Business Management," "Human Resources," "Leadership," "Hotel Law," "Accounting," "Finance," "Marketing," and "Information Systems and International Relations," could develop students' professional understanding of the tourism industry and improve critical thinking skills as well as broaden their vision. At the same time, based on the detailed study of management and business, the courses have the objective of readying graduates for management and leadership roles in the hotel and tourism industries (Song & Yang 2015).

Third, practical training, global student exchange, paid internships, and other courses operationalized in schools, domestic and global hotels, or tourism companies targeting operational service, supply and management play an important role in cultivating students to acquire problem-solving skills, international perspectives, social networks (Shi 2009), team building and leadership skills, all of which are essential in tourism and tourism-related industry management.

Comparison of Chinese and international curriculum setting

Differences in professional training objectives

Talent training is defined by the regulations of higher education which has been influenced by different histories, cultures, values, and concepts. There are differences in the goals between local and international curriculum settings.

Tourism colleges in China have a unified training goal that meets various requirements

In 2012, the Ministry of Education upgraded tourism management to a first-level major under Management discipline and set a goal of "fostering tourism professionals to excel in a management position of tourism administration, enterprise and institution" (Ministry of Education 2012). This goal has set a new direction in knowledge structure, career orientation, and application skills.

The first objective is to master comprehensive knowledge. Every student should master the fundamental theories and skills that meet the requirement of their undergraduate degree and be able to apply the skills in their work environment and conduct entry-level research. Some scholars had defined this goal as "Comprehensive Applied Talents" (Lin & Chen 2008). In other words, students should acquire expertise in their chosen subject and also gain a general understanding of humanity, society, and natural science. Ideally, students should have basic theoretical knowledge and skills in two or more subjects.

The second objective is to fulfill social needs. In the context of national industrial development, combining academic backgrounds and characteristics of tourism colleges, "Global Industrial Leaders" training requires precise classification and accurate labor division. Versatile talents training that can adapt to universal circumstances is also needed to fulfill industry demand.

The third objective is to focus on skills building. Every institute needs to apply advanced educational concepts (based on practical training and leadership building) to enrich students' management skills and build the ability to conduct research (Fan 2011). Course curriculums need to provide the opportunity for students to interact with industry executives and professionals in order to gain practical experience. As a result, students can combine their interests with career professionalism and improve their employability at the same time.

Overseas tourism colleges have diverse training objectives and features of globalization

Overseas schools have educational foundations with well-defined goals that focus on middle and senior management training, such as "cultivating 21st-century industry leaders," measured by a keen sense of business acumen, internationalized thinking, and professional skills.

First is a focus on social services. Foreign colleges often took into account students' emotional intelligence (EQ) alongside professional skills to expand their knowledge and vision, enhancing self-initiative and developing their insight and sensitivity toward social issues and industrial developmental characteristics, resulting in students obtaining social responsibility and judgment.

Second is the trend of globalization. An open and joint curriculum can enrich students' choices and elevate their participation. Some colleges decide to partner with national school-enterprise corporations, respecting students' interests, and refining their knowledge from globalization. Many students gain initiative and enthusiasm from these collaborations, which helps them become a core talent for tourism development with the transcendence of local limitations and international views and problem-solving skills.

Third is the use of professional certification. Applied skills are emphasized in core content, and there is cooperation with enterprises to develop students' professional skills, communication, language, and teamwork. Some colleges adopt vocational certification as one of the graduation criteria, and such certification measures students' skill mastery and help them transition to the workforce successfully, establishing their career goals and meeting industrial standards.

Differences in the course structure

Based on the comparison of the Chinese and foreign curriculum setting above, the curriculum system of Chinese and overseas tourism colleges is summarized as follows.

The curriculum setting in Chinese tourism colleges is homogenized and broad-based coverage

Nowadays, the content of tourism courses in every institute is broadly identical with the course structure meeting the general requirements of the Department of Education in the following areas: (1) integration of various subjects in appropriate proportion. Following the trend of the tourism industry, management as a basic model infiltrates related classes to broaden the application field of theoretical knowledge; (2) the ratio between professional courses and core courses is well distributed. Colleges actively respond to the principles of national higher education to widen the caliber of professionalism, adhere to established morality, and build a holistic and consistent knowledge structure; (3) balance between theory and practice. Under the current "order-based," "zero distance," and other emerging training models, the proportion of practical courses is gradually being adjusted, thereby improving the scientific and pragmatic nature of the course structure and help to achieve the training goals.

Flexible and directed curriculum setting in foreign tourism colleges

Some foreign tourism colleges emphasize professional and systematic theoretical knowledge construction and follow the principle of theory-in-practice, which is targeted and practical and highlights core courses' dominant position. Colleges set management, tourism, and other professional content as independent themes, and interdisciplinary content branch out with clear purpose and guidance. Colleges also emphasize the equilibrium between general education and professional courses. Each college explores the underlying relationship between various subjects and majors, setting the proportion of courses appropriately and differentiating the knowledge structure, impacting industry development. Also, theory and practice can effectively connect and integrate, thereby establishing a more systematic and comprehensive knowledge structure.

Differences in hours/credits

The course credits in Chinese tourism colleges emphasize the key points, with even distribution of hours

The total credits of most Chinese colleges and universities for the tourism major are typically around 150. There is a minimum credit requirement following the standard of credit setting of international courses. The design of course credits is similar in different colleges and universities. Among them, the percentage of credits for compulsory courses, including public courses and specialized courses, is high, basically accounting for 60%–70% of the total credits and highlighting the balance of general courses requirements with some focus on specialized courses. The main compulsory courses have the most credits and the most extended hours, highlighting the focus on the professional courses. The hours of the public courses, specialized theory courses, and practical skill courses are evenly distributed in different grades, progression from synthesis to specialty, from literacy to skills, from theory to application, consistent with the overall training goals.

There is an appropriate balance and flexibility amongst the course credits of overseas tourism colleges

The general education system length of foreign tourism colleges is typically three to four years. According to the curriculum, the credit requirement is around 120 points. Students must achieve a minimum number of credits, and there is no limit on the maximum number of credits. The type and modular credits for required courses and electives are flexible enough to allow independent choice, with students forming a tendency to have a combination of professional and personalized learning. The proportion of practical hours is significant, ranging from 600 to 1,300 hours, highlighting the importance of practicums and internships in the industry. The theoretical courses have formats such as lectures and discussions. Guiding students to learn and think independently is prioritized over theoretical teachings.

Conclusion and discussion

As one of the many strategic mainstay industries of Chinese national economic development, tourism has become widely influential and is proliferating within the country. Due to its diversification and specialization, tourism demands a significant number of professionals with higher requirements. Meanwhile, tourism development has also caused its own set of problems, such as mismatch of education training goals and industry development and the imbalance between tourism talent supply and industry demand in China. Comparing curriculum settings between Chinese and acclaimed foreign tourism colleges has become one solution to this ongoing issue. The course content and course credits, scientific measure of credit hours, practical training setting, the systematic knowledge construction, and the compatibility of skill training and training objectives have comprehensive impact and guidance on training the talents.

This study selects Chinese and overseas well-known tourism colleges as practical examples, proposing the importance of course content, academic subject background, and industry diversity based on course category. This study also emphasizes the combination of theoretical learning and practical application, professional knowledge and general education, and the compatibility between course features and training objectives. Meanwhile, the differences of national conditions have generalized personalization and differentiation characteristics. Accordingly, diversity and distinction of development goals, service scope, curriculum structure between Chinese and overseas talents are distinctly emphasized. Therefore, this study reflects on the homogeneity and consistency on which Chinese tourism colleges are focusing and the dynamics and diversity of foreign college curriculum. Moreover, this study also points out that the curriculum in Chinese tourism education requires improvement to narrow the gap with foreign tourism colleges and industry development.

Above all, this study compares and analyzes the curriculum design between advanced Chinese and foreign tourism colleges as a general guide to Chinese tourism education. Under the influence of the global tourism industry, there are some key development trends:

- Tourism education driven by industry has entered a new era with full throttle speed. The curriculum content of tourism colleges will continuously influence the supply of human resources worldwide, and the curriculum system will continue to be improved and adjusted. A systematic and comprehensive understanding of theoretical and applied skills will become essential for professional experts in the tourism industry.
- During the period of economic and social transformation, occupations have become more refined and comprehensive. Supported by the Chinese educational system, tourism colleges' curriculum setting is more aligned with the professional systematic knowledge framework, highlighting the industry's advantages and ambition and gradually balancing the demand and supply.
- Colleges are paying more attention to building a training model that reflects professional characteristics. Considering international innovations, the curriculum setting of tourism colleges will closely follow the characteristics of the industry, strengthen globalization and industrial collaboration, and establish achievable "teaching standards."

The curriculum setting in tourism colleges is the basis for ensuring talent training goals, an important signifier for the accurate interpretation of industry characteristics and in-depth analysis of knowledge, and the impetus for the compelling connection between talents and enterprises. The booster in demand promotes professional training to fill the gaps of the Chinese tourism industry and facilitates the development of the industry.

References

Bao, J & Zhu, F 2008, 'The problem and outlet for shrinking of Chinese tourism undergraduate education: reflections on 30 years' of tourism higher education development', *Tourism Tribune*, vol. 23, no. 5, pp. 13–17.

Bo, X & Yao, Y 1999, 'The development trend of hotel management education in Western Europe and the development direction of hotel restaurant education in my country', *Journal of Guilin Institute of Tourism*, vol. 10, no. 2, pp. 70–73.

Fan, Y 2011, *Current status and trends of China's tourism talent development*, Tourism Education Press, Beijing.

Feng, Y & Li, Y 2007, 'A comparative study on the undergraduate curriculum system of tourism specialty between China and Japan: taking Nankai University and Lijiao University as examples', *Information Science and Technology*, no. 31, pp. 184+213.

Gerenjianli 2019, *2020 ranking of tourism management major in Chinese university*. Available from: http://www.gerenjianli.com/zypaiming/ff615qa4.htm [27 July 2020].

Gong, J & Chen, C 2018, 'Current situation and counter measures of entrepreneurship education in tourism management majors in universities', *The Guide of Science & Education*, no. 32, pp. 171.

Guo, Q, Ma, Y, & Liu, M 2007, 'Conception of curriculum system reform of college tourism management major', *Journal of Adult Education College of Hubei University*, vol. 25, no. 2, pp. 66–68.

Guo, J & Yang, C 2008, 'Analysis of the structure optimization of tourism management professional university theory courses and practical classes', *The 30-Year Academic Forum of Jiangsu Tourism Development and the Annual Meeting of Jiangsu Tourism Society*, Jiangsu, 19–20 December.

Han, B & Lu, P 2010, 'Comparison of practical teaching of tourism education in general colleges and universities', *Human Geography*, vol. 25, no. 6, pp. 156.

Hong Kong Programmes n.d., *Programme structure*, The Hong Kong Polytechnic University. Available from: https://shtm.polyu.edu.hk/academic-programmes/hong-kong-programmes/bachelor-of-science/bachelor-of-science-bsc-hons-in-hotel-management/programme-structure/ [27 July 2020].

Huang, J 2009, 'Comparative study on tourist educational institutions of higher learning and research cooperation: institute of tourism at Cornell University school of hotel and Beijing United University', *Tourism Tribune*, vol. 24, no. 2, pp. 87–91.

Jiang, X, Wu, J, & Huang, Y 2003, 'Survey and innovative research on the current situation of curriculum system of tourism management major-taking Guilin Institute of technology as an example', *Journal of Guilin Institute of Tourism*, vol. 14, no. 4, pp. 70–76.

KohK, D 1995, 'The four-year tourism management curriculum: a marketing approach', *Journal of travel Research*, vol. 34, no. 1, pp. 68–72.

Li, J & Ji, Z 2007, 'Comparison of tourism education in China and France', *Journal of Inner Mongolia Normal University (Educational Science)*, vol. 20, no. 5, pp. 112–114.

Li, L & Zhang, M 2010, 'A comparative study on the undergraduate curriculum design of tourism management major: taking five universities in Guangzhou as an example', *Forum on Contemporary Education*, no. 9, pp. 63–65.

Lin, G 1998, 'A preliminary exploration of the teaching content and curriculum system of tourism management major in management disciplines', *Tourism Tribune-Tourism Education Magazine*, no. S1, pp. 66–69.

Lin, Y & Chen, W 2008, 'A probe into the undergraduate training objectives of tourism management major and students' employment deviation', *China Science and Technology Information*, no. 14, pp. 200–201.

Liu, T 2003, 'An analysis of the undergraduate education in Chinese and foreign hotel management majors-taking Cornell University hotel college and Beijing Union University tourism college as examples', *Tourism Tribune*, no. S1, pp. 73–76.

Liu, H 2004, 'Problems and strategies faced by undergraduate education of tourism management majors in colleges and universities', *China Higher Education Research*, no. 9, pp. 76–78.

Lu, H 1999, 'Several issues about the undergraduate courses of tourism major in colleges and universities', *Journal of Guilin Institute of Tourism*, no. S2, pp. 140–142.

Luo, Z & Luo, Y 1997, 'Research on the curriculum system design of higher education tourism management specialty', *Tourism Tribune*, no. S1, pp. 58–59.

Ministry of Culture and Tourism of the People's Republic of China 2011, *Chinese tourism industry "twelfth five-year" talents plan (2011–2015)*. Central People's Government. Available from: https://wenku.baidu.com/view/a1f089c158f5f61fb7366625.html [27 July 2020].

Ministry of Education 2012, *Colleges and universities undergraduate course catalog*. Available from: http://www.moe.gov.cn/srcsite/A08/moe_1034/s3882/201209/t20120918_143152.html [27 July 2020].

School of Hotel and Tourism Management n.d., *School of Hotel and Tourism Management Homepage*. Purdue University. Available from: http://www.cfs.purdue.edu [27 July 2020].

Qian, X 2005, 'On the enlightenment of Australian tourism education to my country', *Vocational and Adult Education*, no. 5, pp. 27–28.

Quacquarelli Symonds 2019, *2019 world university tourism management hospitality & leisure management professional ranking*. Available from: http://ranking.promisingedu.com/2019-qs-all-undergraduate-hospitality_leisure_management [27 July 2020].

Shanghai Ranking Consultancy 2020, *Shanghai ranking's global ranking of academic subjects 2020: hospitality & tourism management*. Available from: http://www.shanghairanking.com/Shanghairanking-Subject-Rankings/hospitality-tourism-management.html [27 July 2020].

Shi, L 2009, 'A comparative study on the training modes of Chinese and foreign tourism professional talents', *Internationalization of China Tourism Higher Education Internationalization Summit Forum*, Shanghai, 20–21 June.

Song, H & Yang, H 2015, 'Yang Hui-jun hotel and tourism education to open up a new era: school of hotel and tourism management at The Hong Kong Polytechnic University as an example', *Tourism Tribune*, vol. 30, no. 9, pp. 6–9.

Strohberhn, C 1994, 'Marketing and recruiting efforts as perceived by administrators and students of hospitability graduate programs', *Hospitability and Tourism Educator*, vol. 6, no. 1, pp. 33–37.

Su, J & Yin, H 2005, 'Tourism management introduces "Green Globe 21" research and implementation of modular teaching method', *Journal of Higher Education*, no. 11, pp. 92.

Sun, J 2013, 'Thoughts on several issues of curriculum design of tourism management major in colleges and universities', *New Economy*, no. 29, pp. 110–111.

Sun, J & Li, H 2019, 'On "total" and "special" in the reform of tourism management teaching: based on the design of tourism professionals in 43 universities in the United States', *Humanities World*, no. 6, pp. 7–13.

Tan, B & Mao, F 2001, 'A comparison of the undergraduate teaching systems in tourism majors at home and abroad', *Theory and Practice of Higher Education in Building Material Science*, vol. 20, no. 10, pp. 105–108.

Tong, H & Lu, W 2019, 'Optimization of college tourism management courses based on Industry demand-taking Chongqing Three Gorges University as an example', *Journal of Hubei Open Vocational College*, vol. 32, no. 1, pp. 139–140+143.

Wang, W 2000, 'A comparative study on the undergraduate teaching of tourism management in China and the United States', *Journal of Guilin Institute of Tourism*, vol. 11, no. 1, pp. 5–8.

Wang, C, Hong, Y & Niu, Z 2016, 'Research on the undergraduate major construction of tourism management in universities', *Journal of Jiangsu University of Technology*, vol. 22, no. 6, pp. 102–105.

Wu, B, Tang, Z & Cai, L 2002, 'Tour of American universities', *Tourism Tribune*, vol. 17, no. 5, pp. 76–79.

Xie, Y 2003, 'Tourism and hospitality research: a comparison between China and foreign countries – on the maturity of Chinese tourism discipline', *Tourism Tribune*, vol. 18, no. 5, pp. 20–25.

Xu, C 1999, 'Exploring the curriculum system of undergraduate tourism management major', *Journal of Guilin Institute of Tourism*, no. S2, pp. 143–145.

Xu, H & Zhang, C 2004, 'A comparative analysis of overseas and Chinese tourism education and its enlightenment', *Tourism Tribune*, no. S1, pp. 26–30.

Xu, Y, Xu, X & Wang, J 2016, 'University curriculum construction based on complete credit system', *Education and Vocation*, no. 23, pp. 90–93.

Yuan, S, Meng, T, Miao, F, Zheng, L, He, F & Gao, Y 2005, 'Research on the development of tourism education in colleges and universities', *Tourism Science*, vol. 19, no. 6, p.72–75.

Yue, D & Li, H 2009, 'The reform of the curriculum system of funny tourism management specialty based on the training of interdisciplinary talent and applied talents', *Journal of Xi'an University of Arts & Science (Natural Science Edition)*, vol. 12, no. 2, pp. 119–122.

Zeng, G & Peng, Q 2008, 'A systematic analysis of the undergraduate curriculum system of tourism management major: based on the questionnaire survey of senior undergraduates in top-rated universities', *Journal of Guilin Institute of Tourism*, vol. 19, no. 3, pp. 443–447.

Zhang, D 2015, 'The training mode of applied innovative talents in college tourism management major', *Academic Exploration*, no. 2, pp. 73–77.

Zhao, P & Wang, H 1998, 'Research on the reform of the teaching content and curriculum system of tourism management majors in the 21st century', *Tourism Tribune-Tourism Education Magazine*, no. S1, pp. 21–27.

Zhuang, J 1998, 'A comparison of higher education in tourism between my country and foreign countries', *Tourism Tribune*, no. S1, pp. 47–50.

10 International collaboration in tourism higher education

Qiuju Luo and Xueting Zhai

Introduction

Against the background of China's reform and opening up and globalization, internationalization has become a mainstream trend of higher education in the world (Xia, 2007). Internationalization has also become an effective way for Chinese universities to strengthen opening up and building high-level programs, and is also an important strategy for universities to enhance their international competitiveness. International school operation is an integration of the education system and rules, which can develop students' intellectual potential and capability. Sino-foreign cooperation in higher education management and interschool exchanges are also important ways to introduce advanced overseas higher-education concepts, management systems, teaching excellence, curricula, and textbook systems into China's education system. At the same time, internationalization also promotes the linkage of existing resources of domestic colleges and universities, and finally produces a win-win synergy. More important, international collaboration could serve the development and construction of the country. Under the strategy of the Belt and Road Initiative, universities in China have begun to focus on universities in the "one belt and one way" countries and jointly train the required talents for those universities.

As an academic discipline, tourism has a high degree of openness; thus, it is very important to promote the internationalization of tourism colleges and universities in China. In recent years, Chinese tourism schools have actively explored and developed international cooperation, and the level of international education has significantly improved. On the whole, China's tourism education shows multilevel and multipath characteristics in the Sino-foreign cooperation in higher education management and has initially formed an international training system that includes undergraduate, masters, and doctoral degrees (Huang & Zhang, 2011).

However, further development of international cooperation at this stage still faces difficulties and challenges. The country promotes the "Double

DOI: 10.4324/9781003004363-10

First-Class" initiative of first-class universities and first-class disciplines, emphasizing "bringing in" over "going out", especially focusing on the cultivation of local disciplines and talents. The decline of macro support challenges the development of international cooperation between China and foreign countries. It is of practical significance to further explore the international cooperation between Chinese and foreign tourism education by sorting out and summarizing the international cooperation situation of mainland Chinese tourism institutions.

Sample cooperative programs

Based on the evaluation report of Chinese universities and disciplines jointly issued by the Research Center for Chinese Science Evaluation (RCCSE), the China Education Quality Evaluation Center of Wuhan University, and the China Science and Education Evaluation Network, this chapter takes 20 key universities in tourism management as examples (see Table 10.1) to sort and summarize the general situation of tourism higher education cooperation between China and foreign countries.

Over half of the sampled tourism schools have established teaching cooperation with national/overseas institutions and launched an international cooperation project covering undergraduate, masters' and doctoral education. Data show that the bachelor double degree cooperation (56%) is the most frequently chosen form of international cooperation in mainland Chinese tourism institutions, followed by the "bachelor + master" double degree (23%), which reflects the greater operability and popularity of mainland Chinese cooperation, and then the master's double degree (18%) and the joint PhD program (3%) (see Figure 10.1). There is still much room for international cooperation at the postgraduate level.

Table 10.1 Twenty key universities with Tourism Management programs

University	
Sun Yat-sen University	Huaqiao University
Yunnan University	Hubei University
Beijing International Studies University	Dongbei University of Finance & Economics
Beijing Union University	Xiamen University
Nankai University	Fujian Normal University
Shanghai Normal University	Sichuan University
Tianjin University of Commerce	Northwest University
Jinan University	Fudan University
Guilin University of Technology	Shenyang Normal University
South China University of Technology	Zhengzhou University

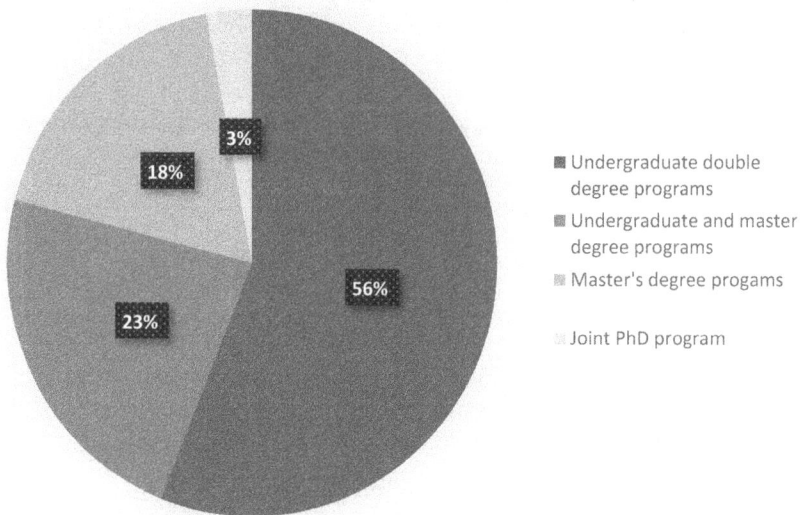

Figure 10.1 International cooperation projects of sampled tourism institutions.

Modes of Sino-foreign cooperation at the school level

Undergraduate double degree cooperation

The "4 + 0" mode

OVERVIEW

The "4 + 0" mode refers to the four-year study in domestic colleges and universities under the training plan jointly formulated by Chinese and foreign universities, and after graduation students can obtain degree certificates from both domestic and foreign universities. Based on the educational resources of domestic schools, this collaboration mode integrates the excellent resources of foreign schools to achieve the goal of training international professionals. Even in China, students can still accept international education and share international resources. The "4 + 0" mode is one of the most strongly supported international cooperation modes.

ADVANTAGES

- Sino-foreign cooperation: Chinese and foreign colleges and universities set up a training system in which students receive international teaching at home.
- Double degree: Double-degree certificates of two universities.
- Moderate tuition fee: Lower than the cost of studying abroad.

PROBLEMS AND CHALLENGES

- Cultivation system: Challenges in setting up and implementing a cultivation system jointly run by Chinese and foreign colleges and universities.
- Degree recognition: The foreign degree may be questioned due to the person having studied in China.

CASE: TUC–FIU COOPERATION COLLEGE IN TIANJIN UNIVERSITY OF COMMERCE

Tianjin University of Commerce (TUC) and Florida International University (FIU) founded a cooperative college in July 2007. It has two majors: Hotel Management (Sino-US cooperation) and Tourism Management (golf operation and management). With the approval of the academic degree committee of the State Council (Academic Degree Office [2004] No. 73), TUC and FIU jointly initiated the Bachelor of Science Education Project of Hotel Management.

COOPERATION MODE

In the professional cooperative project in hotel management (Sino-US cooperation), TUC is responsible for the delivery of the basic courses of years one and two and English language teaching, while FIU is responsible for the teaching and management of the final two years of professional courses. In such an arrangement, the FIU teaching mode is adopted, the original textbook is used, and FIU sends teachers to provide professional courses delivered in English.

To build the college into a domestic first-class and internationally famous hotel management college, TUC invested 220 million yuan for the cooperative project at the beginning and built a modern green campus integrating teaching, demonstration areas, offices, and catering. According to the requirements of FIU, TUC has designed and equipped the college with a first-class Western cuisine kitchen and a Western restaurant, a wine appreciation room, a hotel information technology center, a language lab, and a multimedia classroom. Based on the needs of the international hotel market for management talents, the cooperative project integrates the advantages of China and the United States, draws lessons from the high-quality education resources of the United States, constructs and practices an innovative training model of international hotel management talents that combines quality education with professional education, and integrates theoretical and practical learning. In the design of the talent training program, we should always take the concepts of international tourism enterprises for talents as the guide and build a reasonable knowledge, ability, and quality structure for students. Here, the "one main line, two systems" hotel management personnel training model has been implemented, aiming to build a theoretical teaching system and professional ability training system, cultivating excellent character, basic quality, professional and innovation ability as the main line, and organically combining the acquisition of knowledge, training ability, and improvement of quality.

CURRICULUM

In terms of curriculum, the teachers from TUC are responsible for the teaching of the first two years. Students study the courses of these two years in the headquarters of the university to complete 60 credits, including the compulsory courses required by FIU. Students who have passed the final examination are recognized by both FIU and TUC. The teaching plan for the final two years are provided by FIU, and all the professional courses of TUC are succeeded by the professional courses of FIU. Students need to complete the professional courses required by FIU, acquiring a further 60 credits as well as an internship and the thesis to meet the requirements of the Tourism Management Major of TUC. FIU sends five teachers (with rich teaching and hospitality industry experience) to teach each year. Those students who complete their studies will be awarded bachelor's degrees by both FIU and TUC. The basic courses include English composition, humanities and writing, quantitative reasoning, natural science, art, and social investigation. The comprehensive courses aim to cultivate students' wide-ranging talents (Xia, 2007).

The "3 + 1" mode

OVERVIEW

The "3 + 1" mode refers to students receiving three years of study in a local university and then one year in an international cooperative university. After graduation, students can obtain bachelor's degrees from both domestic and foreign universities. The "3 + 1" joint education and training mode provides a platform for students to study and live overseas for one year, and at the same time, they can obtain double degrees and double diplomas from the local and foreign universities. Compared with the cost of studying abroad, this arrangement is only moderately expensive and more acceptable for ordinary families to allow students to study abroad.

ADVANTAGES

- Integration of Chinese and foreign country expertise: Combination of professional learning at home and abroad and international integration.
- Overseas experience: One-year overseas study and life opportunities.
- Double degree: Double degree certificates from two universities, one at home and one abroad.
- Moderate tuition: Students from ordinary families can also get the opportunity to study abroad.

PROBLEMS AND CHALLENGES

- Teaching connection: Chinese and foreign universities need to connect in the setting and learning of professional courses, which offers challenges to teaching cooperation.

- Credit certification: There are difficulties in the methods and certification of double degrees that need to be supported by both schools.
- Study and life: Challenges for students to adapt to foreign study and life in only one year.
- Language difficulties: Learning French is a challenge for Chinese students.

CASE: TOURISM MANAGEMENT PROJECT OF SCHOOL OF TOURISM MANAGEMENT, SUN YAT-SEN UNIVERSITY, AND THE UNIVERSITY OF ANGERS (FRANCE)

The "3 + 1" Tourism Management Project of the School of Tourism Management of Sun Yat-sen University (SYSU) in collaboration with the University of Angers, France, was launched in 2012. It is a double bachelor Diploma Program of China-France Tourism Management jointly initiated by SYSU and the University of Angers. The program enrolls 35 Chinese students every year. Students who successfully pass the baccalaureate examinations of both institutions and meet the graduation requirements of both can obtain a bachelor's degree in management from SYSU and a bachelor's degree in tourism, heritage, hotel, catering, and exhibition in social sciences from the University of Angers.

To date, the project has enrolled a total of 142 students, including 25 students in 2012, 24 in 2013, 25 in 2014, 35 in 2015, and 33 in 2016. Of the first graduates, 60% continued on to study for a master's degree. Some students have been admitted by European and American universities, such as the Paris Higher Business School, London Business School, Duke University, and Lille University for their outstanding achievements and have obtained scholarships. Some students remained to study or work in France and other European countries.

TEACHING DEVELOPMENT

The two universities have carried out in-depth cooperation in teaching and have actively formulated a specialized syllabus for the education stages of a double bachelor's degree recognized by both China and France. To better realize the joint training shared by China and France, the School of Tourism Management, SYSU, focuses on strengthening French teaching so that students' French language level can meet the requirements of the lectures and exchange activities of the University of Angers. In addition, teachers from the University of Angers deliver eight courses in Zhuhai.

The double bachelor's degree program of the School of Tourism Management, SYSU, is for four years, eight semesters in total. Students of the program who have obtained the TCF B2 level on the French proficiency test in SYSU in the first three years can attend the College of Higher Tourism Education, University of Angers, to study as senior students. During their study in China, students will complete the teaching plan required by Sun Yat-Sen University and strengthen their French language and cultural learning so that they can continue their studies at the University of Angers. To further enable the students of SYSU to complete their bachelor's degree at the University of Angers, they undertake eight French language courses in Zhuhai, China, including French oral communication, reading, and written expression skills, introduction to the global tourism industry, professional tourism (food, wine, exhibition, hotel, catering, etc.), and general analysis of tourist attractions (tourism attraction analysis, tourism practice, rural and urban tourism, coastal tourism, mountain tourism), writing training, and tourism marketing (tourism consumer behavior, etc.).

However, Chinese students who do not reach the TCF/TEF B2 level in the French proficiency test will not be able to go to France to study in the fourth stage or obtain a diploma from the University of Angers. Instead, they will be enrolled in the general class of the Tourism Management major at SYSU. If they meet the graduation requirements of SYSU, they will get a bachelor's degree from SYSU.

FURTHER STUDY

The graduates of SYSU who have passed the assessment of the fourth academic year in France and have obtained a bachelor's degree from the University of Angers can continue studying by registering for the master's degree at Angers. At present, two students, having obtained double bachelor's degrees, have chosen to continue their graduate studies at the University of Angers.

The "2 + 2" mode

OVERVIEW

The "2 + 2" mode refers to studying in a domestic university for the first two years and then studying in an overseas cooperative university for the last two years. After graduation, students can obtain bachelor's degrees from two universities at home and abroad. This mode combines the current teaching ideas and experience at home and abroad, which is conducive to the cultivation of international, professional, and high-level international talents.

ADVANTAGES

- Teaching at home and abroad: Fully integrating advanced teaching concepts and experience at home and abroad and cultivating international professionals.
- International horizon: Two years of study and life overseas, broadening the international vision of students.
- Double degree: Double degree certificates from two universities, one at home and the other abroad.

PROBLEMS AND CHALLENGES

- Curriculum: Challenges the close cooperation in curriculum and teaching.
- Credit certification: The issuance and certification of double degrees need the support of both schools, and there may be difficulties.
- Cultural differences: Students may face issues adjusting to study or life abroad.

CASE: EXHIBITION ECONOMY AND MANAGEMENT PROJECT IN THE
SCHOOL OF TOURISM MANAGEMENT, SUN YAT-SEN UNIVERSITY, AND
THE UNIVERSITY OF QUEENSLAND (AUSTRALIA)

In 2009, the School of Tourism Management, SYSU, and the University of Queensland (UQ) set up a "2 + 2" joint training program in the majors of Exhibition Economy and Management, which enrolls 40 students every year. It aims to introduce the advanced teaching experience of the UQ, give full play to the advantages of teaching and research resources of both cultures to cultivate excellent talents in economy and management featuring professional quality, an international view, and practical experience. Students who have passed the International English Language Testing System (IELTS) language test in the first three semesters are eligible to continue studying in UQ in the third and fourth semesters. The core courses adopt the independent small class teaching mode in either an English or bilingual setting. The teaching staff are all foreign teachers with overseas study background. In addition, teachers at UQ are invited to teach courses on the Zhuhai campus of SYSU on a regular basis. After graduation, students will receive a bachelor's degree in management from SYSU and a bachelor's degree in international hotel and tourism management from UQ. At present, the program has successfully trained 308 students, about 70% of whom have chosen to continue their studies in universities around the world after graduation.

CURRICULUM

All courses of "2 + 2" cooperative education are jointly discussed and completed by the professional teachers at the School of Tourism Management, SYSU, and UQ. Based on the small class teaching mode, the basic theory with a professional in-depth method is adopted to ensure a better connection and transition between the courses of exhibition economy and management at the two universities. Students can adapt to the English teaching mode and method in Australia in advance. Every year, the, Sun Yat-sen's School of Tourism Management sends two teachers to the UQ to follow up and revise the project agreement and visit the "2 + 2" students studying in Australia.

In terms of teaching, all teaching materials of both the professional compulsory courses and the professional elective courses use the original English textbooks. In addition, the teaching staff are teachers with overseas study backgrounds or are foreign teachers. More than 95% of all the teachers in the School of Tourism Management have experience in studying overseas, including the United States, Japan, the Netherlands, Australia, Germany, and Hong Kong Special Administrative Region (SAR).

TALENT CULTIVATION

Those "2 + 2" course students who have successfully visited Australia have performed well at the UQ, and 70% of the students have chosen to continue their studies at famous foreign universities. Some of the outstanding graduates have received full scholarships to pursue their doctorate after completing their honors degree at UQ. In the Dean's Honour Roll in 2015, five students won the title of Outstanding Graduates in the major of International Hotel and Tourism Management. At the graduation ceremony, Lizi Chen, one of this cohort's students, was designated as the student representative and spoke on behalf of the graduates from the Business School.

In terms of employment, graduates mainly work in foreign-funded enterprises, educational institutions, and exhibition companies, engaging in planning and design, marketing, public relations, operation management, and budget evaluation. Most of the students return to work in China or Hong Kong SAR, while a few choose to work abroad.

Bachelor + master double degree cooperation

The "4 + 1" mode

OVERVIEW

In the "4 + 1," after four years of undergraduate study in a domestic university and obtaining a bachelor's degree, students will study for a master's degree in a foreign cooperative setting in the fifth year.

ADVANTAGES

- Learning at home and abroad: Fully integrated advanced teaching concepts and experience at home and abroad cultivating international professionals.
- International vision: One-year study and life overseas, broadening the international vision of students.
- Double degree: Bachelor's degree from a domestic university and a master's degree from a foreign university.

PROBLEMS AND CHALLENGES

- Time pressure: Students need to complete the course and the master's thesis during the one-year master's degree study.
- Study arrangement: Students need to complete the graduation thesis in the fourth year of the undergraduate course and study in the master's course or advanced undergraduate course at the same time, which is challenging.

- Cultural differences: Students may face difficulties adjusting to study or life abroad.

CASE: THE "4 + 1" MASTER'S DEGREE PROGRAM OF THE SCHOOL
OF TOURISM MANAGEMENT, SUN YAT-SEN UNIVERSITY, AND THE
UNIVERSITY OF SURREY (UK)

After four years of undergraduate study and obtaining a bachelor's degree in the School of Tourism Management, SYSU, the students go to the University of Surrey in the UK for a one-year postgraduate course in the fifth year. Upon successful completion of the graduate program at the University of Surrey, students will receive a master's degree from that university. The School of Hotel and Tourism Management, the School of Business, and the School of Economics of Surrey University will offer master's program courses to students who meet the selection conditions, and students can choose their majors and courses according to their own interests and career goals.

This master's degree program provides students with world-class education standards and improves their employment opportunities and career choices in the global market. Advantages include:

- Opportunity to study in the UK, improve language skills and self-confidence, and have the opportunity to establish a global social network.
- Have a challenging and fruitful learning experience, broaden the knowledge field in a short time, and obtain a master's degree from a famous international university.
- First-class teaching and a high employment rate after graduation.
- The UK education system adopts interactive and challenging teaching styles, develops students' independence, leadership, communication skills, and critical thinking ability, and lays a solid foundation for students to embark on their career journey.

The "3 + 2" mode

OVERVIEW

The "3 + 2" mode means that students study for a bachelor's degree at a domestic university for the first three years and then a master's degree at an overseas cooperative university in the next two years. Students who have completed the graduation requirements of both institutions can obtain a bachelor's degree from the domestic institution and a master's degree from the foreign institution.

ADVANTAGES

- Learning at home and abroad: Fully integrated advanced teaching concepts and experience at home and abroad, cultivating international professionals.

- International vision: Two years of study and life overseas, broadening the international vision of students.
- Double degree: Bachelor's degree certificate of the domestic university and master's degree certificate of the foreign university.

PROBLEMS AND CHALLENGES

- Curriculum: Challenges the close cooperation between curriculum and teaching.
- Learning arrangement: Students need to complete their graduation thesis in the fourth year and study for a master's or take higher-level undergraduate courses, which is challenging.
- Cultural differences: Students may face difficulties adjusting to study and life abroad.

CASE: THE "3 + 2" BACHELOR AND MASTER'S DEGREE PROGRAM OF SUN YAT-SEN UNIVERSITY AND TEMPLE UNIVERSITY (US)

This program is composed of two stages: the first three years of undergraduate study at SYSU and the second two years of undergraduate higher-level courses and master's degree courses at Temple University. Students participating in this program will study for a master's degree in Sports Business or in Tourism and Hospitality at Temple University.

Master's double-degree cooperation

The "1 + 1" or the "1 + 1 + 1" mode

In the master's double degree program, graduate students at a domestic university study in our school in the first year and then study overseas at a cooperative university in the second year. If the length of schooling at the domestic university is three years, the students will return to the domestic university in the third year to study and complete their graduation thesis. The schools on both sides will conduct a mutual credit recognition process. Qualified students will receive master's degrees from both the domestic and the overseas universities.

ADVANTAGES

- Learning will be in both Chinese and a foreign language: Teaching will fully integrate advanced teaching concepts and experience in China and abroad, cultivating international professionals.
- International perspective: Studying abroad for one year during the post-graduate period helps students to expand their international perspective and cultivate diversified academic thinking.

- Double degree: Double master's degrees will be awarded by the domestic and the foreign universities.
- Further study: Studying abroad increases the possibility of continuing to study for a doctoral degree abroad.

PROBLEMS AND CHALLENGES

- Curriculum: Challenges the close and unified cooperation between curriculum and teaching.
- Mutual recognition of credit: The research direction and curriculum setting of both sides are less flexible and the mutual recognition of credit is relatively difficult.
- Cultural differences: Students may face difficulties adjusting to study and life abroad.

CASE: DOUBLE DEGREE PROGRAM OF MTA/MBA OF THE DEPARTMENT
OF TOURISM, FUDAN UNIVERSITY, AND FRANCE BUSINESS SCHOOL

Candidates need to pass the Fudan University Master of Tourism Management (MTA) entrance examination to qualify for the program. The project period is two and a half years. In principle, students spend three semesters at Fudan University and two semesters at France Business School (FBS). In the first, second, and fifth semesters, students will study at Fudan University, and during the third and fourth semesters they will study at FBS. Both sides will mutually recognize credits. Students should complete the required courses, obtain the required credits, and pass the examinations at both schools. Then they should submit the research report required by FBS and complete the degree thesis required by Fudan University. Those who pass the graduation examination can obtain the MTA from Fudan University and the Master of Business Administration (MBA) from FBS at the same time. Through two and a half years of systematic training, the purpose is to make students experience different cultural backgrounds and learning environments and become high-quality tourism management talents with an international vision (The Department of Tourism, 2014).

Joint PhD program

Overview

For the joint PhD, two or more Chinese and foreign cooperative institutions jointly train doctoral students. In principle, in the first year of joint training, PhD students will study in their domestic institutions, taking basic and compulsory courses. In the second year, they will start to study in a foreign cooperative institution and begin to carry out dissertation work. Generally speaking, the study period is three to four years. A double tutor system will be implemented, with each of the partner institutions assigning a professor to guide the doctoral students. The cooperation between the two institutions will grant doctoral degrees to doctoral students. While the cooperation between more than two institutions will not encourage the granting of multi-university degrees, the local university will grant doctoral degrees and the cooperative institutions will issue certificates of completion.

Case: joint training PhD program of School of Tourism Management of Sun Yat-sen University, the University of Surrey, and the University of Queensland

To integrate resources and deepen cooperation, the School of Tourism Management of SYSU, the UQ (Australia), and the University of Surrey (UK) are promoting further cooperation in scientific research by carrying out a joint training cooperation between tripartite scientific researchers and PhD students. The three parties support international research cooperation in the field of tourism, hotel, and exhibition, jointly guiding research, holding international scientific seminars, and arranging scholar visits to build an international academic exchange platform. The aims are to promote the research of global frontier issues in tourism, become a research highland in the world, and promote innovation in the field of tourism, hotel, and exhibition; to set up a global research elite center to cultivate a group of high-quality academic talents with international vision; and to build reserve strength for the future development of the tourism industry.

The tripartite joint PhD program aims to establish a platform for international academic collaboration, to promote a global cutting-edge research agenda, and to establish future research priorities based on key global problems. The tripartite initiative further promotes the effective collaboration of researchers at all levels (doctoral candidates and postdoc and senior researchers), produces high-quality collaborative research, and develops international financial cooperation to promote global research activities. The joint training of doctoral students by the three universities is the first initiative of the existing doctoral program of the School of Tourism at SYSU. The three universities are further broadening the international vision of academic research talents, giving full play to the resource advantages of each university, and jointly promoting the establishment of an elite global research center and a platform for international academic mobility.

Overall challenges of Sino-foreign cooperative programs

International cooperation faces a number of serious problems and challenges (see Figure 10.2).

Social-level challenges

The "Double-First Class" universities and disciplines emphasize running schools with Chinese characteristics and strengthening the development of local schools and majors. The support for "3 + 1," "2 + 2," and other international school

Problems and Challenges in International Collaboration

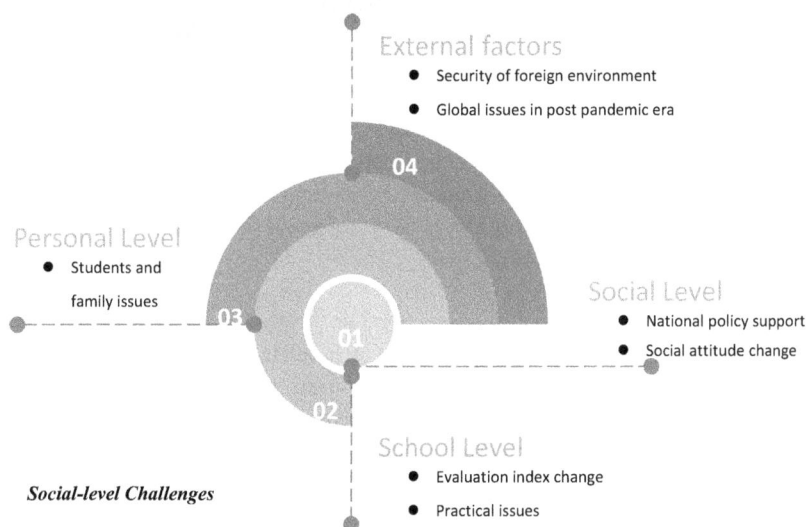

Figure 10.2 International cooperation issues and challenges.

running projects has been weakened, but the "4 + 0" mode is more encouraged, that is, the introduction to China of excellent school operation ideas from abroad as well as resources and experience and the promotion of a domestic discipline structure. This has increased the difficulty and challenge for the expansion of diversified international cooperative education projects.

Nowadays, Chinese society as a whole holds a more rational understanding and attitude towards studying abroad, and people have a new understanding of internationalization. Studying abroad and returning to China is no longer blindly respected but is focused more on the pursuit of quality. For studying abroad, we pay special attention to the choice of world-famous universities, which also puts forward higher requirements for the quality of international cooperation.

School-level issues

In the past, internationalization was an important index to evaluate the quality of a school, but now the evaluation of the rate of going to graduate school is more important than internationalization. The change of the evaluation standard reduces the enthusiasm of the school management in the aspect of internationalization and also causes difficulties for the development of international cooperation.

The practical problems are mainly reflected in two aspects, namely credit exchange and school system differences. In terms of credit exchange, there are some problems in the transfer of credits due to lack of equivalence between the curriculum and training programs of Chinese and foreign universities. If the courses that students take in foreign colleges and universities cannot be seen as corresponding with the courses in the domestic college training plan, the credits they have taken abroad cannot be transferred back. In this case, students need to reselect the courses with missing credits and retake them in domestic schools. As a result, a delay in their graduation date may even take place due to incomplete credits. In terms of school system differences, because of the differences between Chinese and foreign school systems, foreign universities usually transfer students' scores to China in September, while domestic universities have graduated in June, resulting in a likely delay in a student's graduation as the credits were not transferred in time.

Personal-level difficulties

Studying abroad removes students from their familiar environment and comfort zone. They are faced with all kinds of problems in studying and living alone, which challenges their independence and self-management ability. Besides, the teaching environment in foreign countries is relatively relaxed, free, and less restrictive, which also requires students to manage their life and their time more carefully.

While studying abroad, students may find it difficult adjusting to the cultural differences between China and the West. When they go abroad to study and live, many of them will face psychological problems caused by these cultural differences. Introverted students especially are more likely to have these problems, such as being unable to adapt to a foreign environment and integrate into the local life. Therefore, the cooperative universities, both domestic and foreign, should strengthen their support for the students. First, they should adopt a one-to-one buddy system, whereby a mentor can lead the Chinese students to understand and integrate into foreign life. Second, they should pay frequent attention to the overseas students and understand their living conditions and psychological dynamics in time.

In terms of teaching, the cultural differences between China and the West are also reflected in the classroom teaching style. From childhood, Chinese students have been receiving an exam-oriented education. They mainly listen in class and generally lack the ability to interact and think critically. The Western classroom is more open, encouraging students to ask questions and interact, emphasizing critical thinking abilities. On the one hand, this kind of classroom difference can stimulate Chinese students' creative thinking, but on the other hand, it also challenges our students' adaptability.

External factors affecting international cooperation

In 2017, the kidnapping and murder of Zhang Yingying, a Chinese student studying at a US university, raised the whole society's concern about the safety of its overseas students. In a strange environment, there are certain risks to the personal safety of students, which deserves the attention of domestic and foreign cooperative universities alike. We should strengthen the safety aspect for students going abroad, improve their safety awareness, and avoid any possible risks to the greatest extent.

Now, as we enter the post-COVID-19 pandemic era, how can we maintain and strengthen international collaboration of higher tourism education? This question is still a great challenge for both Chinese and foreign universities at a time when it can be difficult to travel from one country to another.

Conclusion

Nowadays, the attitude of the mainland China to international teaching and cooperation is more rational. Instead of blindly seeking internationalization and looking abroad, the teaching system is more objective and the requirements and standards for international cooperation are more stringent. These changes have brought challenges to the further expansion of international cooperation of tourism colleges and universities. With the improvement of the construction of the Chinese tourism infrastructure and the enhancement of discipline strength and influence, the international cooperation of Chinese tourism universities has been upgraded in terms of quality and standards, and the level of international cooperation among different levels of tourism universities has been differentiated. As a result, high-quality colleges and universities will obtain more and better resources, have more space to expand their international cooperation, and raise even higher the quality and level of international cooperation. Meanwhile, general colleges and universities will prefer to seek skilled international cooperation, including international practice and internship. Therefore, for future agreements between Chinese and foreign tourism colleges and universities, it is very important to identify different levels of cooperation and achieve precise demand docking for such cooperation to succeed (Figure 10.3).

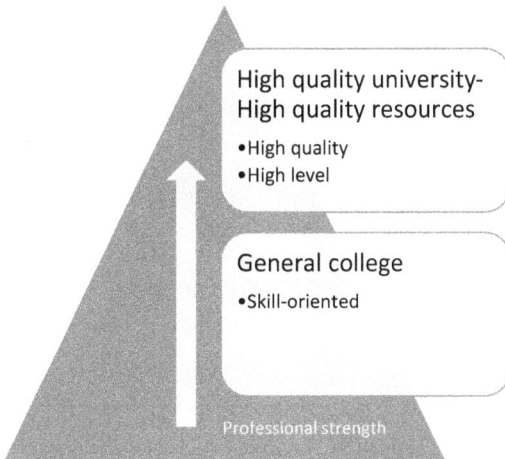

Figure 10.3 Levels of international cooperation in tourism higher education.

References

Huang, F & Zhang, CM, 2011. A study of students' programs of international exchange in colleges and universities, *Journal of Guangdong University of. Petrochemical Technology*, vol. 21, no. 2, pp. 34–36.

The Department of Tourism, 2014. *Fudan University Signed MTA/MBA Double Degree Cooperation Project Agreement with France Business School.* Available from: http://tourism.fudan.edu.cn/17/08/c6623a71432/page.htm. [1 July 2020].

Xia, DX, 2007. *On the pattern and optimization of China and foreign cooperation in tourism education.* Master's thesis. Liaoning Normal University. Available from: CNKI. [1 July 2020].

11 Tourism research in China

Jigang Bao, Ganghua Chen and Songshan (Sam) Huang

Introduction

Over the past few decades, tourism in China has witnessed remarkable growth (Ministry of Culture and Tourism, 2020) and has thus been a significant context and subject area for tourism research (Bao, Chen, and Ma, 2014; Bao, Chen, and Jin, 2018; Bao, Huang, and Chen, 2019; Huang and Chen, 2016a, 2020; Huang et al., 2019; Zhang, Han, Zhang, Zhou, and Zhang, 2019). Chen and Huang (2020) used the term *tourism research in China* to refer to 'the tourism research phenomena in the territorial boundary of mainland China, which include but not are limited to institutions, systems and organizations, policies and research community in relation to tourism research' (p. 365). This term, with its above connotations, is also used in the chapter. Furthermore, it should be noted that due to historical and linguistic reasons and for consideration of data availability, we also refer 'China' in the term 'tourism research in China' to the Chinese mainland, while tourism research in China's other three territories (i.e. Hong Kong, Macau, and Taiwan) is not included in our analysis and discussion.

Previous review articles regarding tourism research in China have examined the development stages (e.g., Bao and Ma, 2011; Bao et al., 2014, 2019; Huang and Chen, 2020), research topics (e.g., Bao et al., 2014; Huang and Chen, 2020), research methods (Huang and Chen, 2016a; Huang et al., 2019), research institutions (e.g., Zhang et al., 2019), and researchers (Yang and Xu, 2017; Zhang et al., 2019) as well as the disciplinary development (see Bao and Lai, 2016). Particularly, with regard to tourism disciplinary development in China, many efforts have been made in the following two aspects. First, issues and challenges faced with upgrading *Tourism Management* to a first-level discipline in China's current disciplinary catalogue (e.g., Bao and Lai, 2016; Chen, 2019; Ma, 2016) have been identified and discussed. Second, the current status of tourism-related disciplines in countries and regions with developed levels of tourism research and education, such as Australia (Huang, 2019; Huang and Chen, 2016b), France (Shen and Philippe, 2019), Hong Kong Special Administrative Region (SAR) (Huang and Chen, 2016b), South Korea (Zhang and Xu, 2017), the UK (Zhang and Chen, 2019), the US (Huang and Chen, 2016b; Yang and Mao, 2019), and the Taiwan region (Wu, Wang, and Lin, 2019), have been introduced to the Chinese mainland.

DOI: 10.4324/9781003004363-11

Despite the commonly accepted viewpoint that tourism research involves multiple source disciplines, that is, geography, management, sociology, anthropology, economics, philosophy, and psychology (e.g., Weiler, Moyle, and McLennan, 2012; Xiao and Smith, 2006), surprisingly few studies have investigated the disciplinary backgrounds of tourism research in China, with a few exceptions (Chen, 2004; Huang, 2011; Lu et al., 2016; Tang, 2013). Although these review studies have undoubtedly advanced our understanding of the disciplinary backgrounds of tourism research in China, they are limited in the following two aspects. First, they used data before 2014 (i.e., 1989–2003, see Chen, 2004; 1999–2009, see Huang, 2011; 2001–2010, see Tang, 2013; 1999–2014, see Lu et al., 2016). Second, they used the China Doctoral Dissertations Full-text Database (CDDFD), which is affiliated with China's largest academic knowledge database (i.e., China National Knowledge Infrastructure [CNKI]), as a data source, neglecting those doctoral dissertations that were not submitted to the CDDFD. Therefore, a comprehensive and up-to-date account of tourism-related doctoral dissertations is needed.

In addition, although institutions and researchers in relation to tourism research in China have been reviewed in previous studies (e.g., Huang and Chen, 2016a, 2020; Zhang et al., 2019), such information seemed to be intermittently and sporadically presented. Therefore, an up-to-date and comprehensive presentation of the recent developments in these aspects is still needed for a timely understanding of tourism research in China.

As such, this chapter aims to present an overview of tourism research in China, including disciplinary backgrounds, research institutions, and researchers. In the following sections, both the data collected by the authors (i.e., the data of doctoral dissertations) and previous literature on tourism research in China will be used to illustrate the evolution and current status regarding tourism research in China.

Disciplinary backgrounds

An overview

It has been widely accepted that tourism research involves multiple source disciplines, that is, geography, management, sociology, anthropology, economics, philosophy, and psychology (e.g., Weiler et al., 2012; Xiao and Smith, 2006). This is also the case to tourism research in China as reflected by doctoral dissertation studies. It is noteworthy that using doctoral dissertations as a data source, it has two aspects of consideration. First, from a realistic perspective, in China, only doctoral dissertations are required to specifically indicate their disciplinary affiliations. Second, the importance of doctoral dissertations in developing tourism research, cultivating the next generation of researchers, and illustrating its cutting edge trends have been widely recognized in previous studies all over the world (e.g., Afifi, 2013; Bao, 2002; Huang, 2011).

Retrievals were conducted separately by using 14 terms equivalent to 'guesthouse', 'holiday', 'hotel', 'leisure', 'travel', 'tourism', 'tourist', 'tourist attractions', and 'visitor' in Chinese as keywords in title. Doctoral dissertations with titles containing keywords equivalent to tourism, holiday, travel, tourist, tourist attractions, vacation, and visitor in Chinese were preliminarily classified as tourism-related dissertations; dissertations with titles containing keywords equivalent to hotel and guesthouse in Chinese were preliminarily classified as hotel-related; and dissertations with titles containing keywords equivalent to leisure in Chinese were preliminarily classified as leisure-related. The search was performed in CDDFD and databases in the libraries of Fudan University, Peking University, Sun Yat-sen University, and Xiamen University. To the best knowledge of the authors, these four universities have doctoral programs in Tourism Management and related disciplines, but did not share their doctoral dissertations to the CDDFD. All doctoral dissertations submitted during 2010 and 2019 were retrieved and were later analyzed. After eliminating those unqualified doctoral dissertations, there were 1,033 doctoral dissertations retained for subsequent analysis.

Among the 1,033 doctoral dissertations, the majority of them (888 dissertations; 86%) are closely related to tourism (i.e., holiday, travel, tourism, tourist, tourist attractions, and visitor), and 9.6% of them (99 dissertations) are focused on leisure, and the remaining 4.4% are concerned with hotel issues (i.e., hotel and guesthouse).

The diversified disciplinary backgrounds of doctoral studies in the fields of tourism, leisure, and hotel in Chinese higher education institutions are shown in Tables 11.1–11.3. As shown in Table 11.1, *Management* (including four first-level disciplines) is the major disciplinary cluster (451/50.8%) contributing to the study of 'tourism' in the doctoral level, followed by *Science* (164/18.5%), and *Economics* (74/8.3%). With regard to leisure-related doctoral dissertations, as shown in Table 11.2, *Management* is still a most contributive disciplinary cluster (21/21.2%), followed by *Education* (17/17.2%), *Philosophy* (15/15.2%), and *Law* (12/12.1%). Similarly, the dominant disciplinary cluster in China's hotel-related doctoral studies is still *Management* (40/87%) (Table 11.3). From the above discussion, it is not difficult to understand that the broad-sense tourism research in China as reflected in doctoral studies has been dominant in the management discipline in the past decade (2010–2019).

Tourism doctoral studies

A closer look into those doctoral studies reveals more specific disciplinary backgrounds of China's doctoral studies in the fields of tourism, leisure, and hotel. As indicated in Table 11.4, among the dominant *Management* disciplinary cluster (451), the *most contributive first-level discipline* to tourism doctoral dissertations is *Business Administration* (366), within which the most contributive second-level discipline is *Tourism Management* (297). This indicates the great importance of a specialized second-level discipline (i.e., *Tourism Management*) to the development of China's tourism doctoral studies and overall tourism studies. The second most contributive first-level discipline within *Management* is *Management Science and Engineering* (45), followed by *Economic Management of Agriculture and Forestry* (28) and *Public Management* (12).

Table 11.1 The disciplinary clusters of tourism-related
doctoral dissertations in China

Disciplinary cluster	Number	Percent (%)
Management	451	50.8
Science	164	18.5
Economics	74	8.3
Engineering	49	5.5
History	42	4.7
Law	38	4.3
Agriculture	22	2.5
Literature	19	2.1
Education	13	1.5
Arts	8	0.9
Medicine	4	0.5
Philosophy	4	0.5
Total	888	100.0

Note: Percentages were rounded up to one decimal point; therefore,
percentages may not add up to 100.0 because of rounding errors.

Table 11.2 The discipline clusters of leisure-related doctoral
dissertations in China

Disciplinary cluster	Number	Percent (%)
Management	21	21.2
Education	17	17.2
Philosophy	15	15.2
Law	12	12.1
Agriculture	10	10.1
Science	9	9.1
Economics	5	5.1
Engineering	4	4.0
History	3	3.0
Literature	3	3.0
Total	99	100.0

Note: Percentages were rounded up to one decimal point; therefore,
percentages may not add up to 100.0 because of rounding errors.

Table 11.3 The disciplinary clusters of hotel-related doctoral
dissertations in China

Disciplinary cluster	Number	Percent (%)
Management	40	87.0
Science	2	4.3
Economics	1	2.2
Engineering	1	2.2
History	1	2.2
Law	1	2.2
Total	46	100.0

Note: Percentages were rounded up to one decimal point; therefore,
percentages may not add up to 100.0 because of rounding errors.

Among the second most contributive disciplinary cluster of *Science* (164), *Geography* (128) has been playing a leading role, with three leading second-level disciplines, namely *Human Geography* (102), *Tourism Geography and Tourism Planning* (10), and *Physical Geography* (9). It is worth noting that *Tourism Geography and Tourism Planning* as a second-level discipline was uniquely and autonomously established in Nanjing University. *Ecology* (24) and *Geology* (9) each also played a critical role in producing tourism-related doctoral dissertations, with most of them completed in those institutions specializing in ecology and geology, such as Central South University of Forestry Science and Technology and China University of Geosciences.

As the third most contributive disciplinary cluster, *Economics* (74) has two first-level disciplines, *Applied Economics* (40) and *Theoretical Economics* (34), both contributing significantly to China's tourism-related doctoral dissertations. Most of the dissertations were completed with a background in *Industrial Economics* (19), *Regional Economics* (12), *World Economy* (16), and *Political Economics* (8), among others.

Leisure doctoral studies

In China, leisure doctoral studies show a slightly different landscape of disciplinary backgrounds. As shown in Table 11.5, the most influential disciplinary cluster is still *Management* (21), with *Business Administration* (10) being its most contributive first-level discipline and *Tourism Management* (7) being its most contributive second-level discipline. It should be noted that in some Chinese universities, for instance, Dongbei University of Finance and Economics, Fudan University, and Zhejiang University, it is acceptable and a common practice that doctoral candidates enrolled in *Tourism Management* choose a leisure topic for their doctoral dissertations.

What is noteworthy is that, slightly different from the distribution identified in tourism doctoral studies (Table 11.4), *Education* (17) becomes the second contributive disciplinary cluster to leisure doctoral studies, especially with the *Physical Education* (16) being a first-level discipline to accommodate sports-related doctoral studies. In addition, *Philosophy* (15) has greatly facilitated leisure doctoral studies in China by contributing eight leisure-focused dissertations within *Leisure Studies*, four dissertations within *Aesthetics*, and three within other second-level disciplinary backgrounds.

Hotel doctoral studies

With a smaller volume compared with that of tourism doctoral studies, doctoral studies in the field of hotel represent a management-dominated disciplinary landscape. Specifically, as shown in Table 11.6, 40 out of the 46 hotel doctoral dissertations were completed within the disciplinary cluster of Management, and notably, 34 of them were within Business Administration and the other six within Management Science and Engineering. Within Business Administration, similar to the patterns in doctoral studies of leisure and tourism, Tourism Management

Table 11.4 The major disciplinary subcategories of tourism-related doctoral dissertations in China

Disciplinary cluster	First-level disciplines (no. of dissertations)	Second-level disciplines (no. of dissertations)
Management (451)	Business Administration (366)[a]	Tourism Management (297)
		Enterprise Management (36)
		Economics and Management of Technology (15)
		Marketing Management (4)/Marketing (2)
	Management Science and Engineering (45)[b]	Management Science and Engineering (44)
		Industrial Engineering and Management Engineering (1)
	Economic Management of Agriculture and Forestry (28)[c]	Economic Management of Forestry (13)
		Economic Management of Agriculture (12)
		Economic Management of Agriculture Fishery (1)
	Public Management (12)[d]	Land Resource Management (4)
		Public Administration (4)
		Organization and Management of Public Works (1); Rural Development and Management (1); Public Economic Management (1)
Science (164)	Geography (128)	Human Geography (102)
		Tourism Geography and Tourism Planning (10)
		Physical Geography (9)
		Urban and Regional Planning (3)
		Cartography and Geographical Information System (1)
		Historical Geography (1); Natural Disaster Science (1); Resource Science (1)
	Ecology (24)	N.A.[e]
	Geology (9)	Quaternary Geology (5)
		Ecological Geology (2)
		Paleontology and Stratigraphy (1)
		Land and Resources Information Engineering (1)
	Decision Science (1); Mathematics (1); Statistics (1)	N.A.[e]
Economics (74)	Applied Economics (40)[c]	Industrial Economics (19)
		Regional Economics (12)
		International Trade (3)
		Quantitative Economics (2)
		Finance (1); National Economics (1)
	Theoretical Economics (34)[d]	World Economy (16)
		Political Economics (8)
		Economics of Population, Resources and Environment (6)
		Western Economics (3)

Engineering (49)	Geological Resources and Geological Engineering (13)	Industrial Economics of Resources (8)
		Earth Exploration and Information Technology (3)
		Tourism Geology and Geological Relics (2)
	Architecture (8)[e]	Architectural Design and Theory (3)
		Landscape Engineering (2)
	Transportation Engineering (5)	Transportation Planning and Management (4)
		Vehicle Application Engineering (1)
	Environmental Science and Engineering (5)	Environmental Planning and Management (2)
		Environmental Science (2)
		Environmental Science and Engineering (1)
	Computer Science and Technology (4)	Computer Science and Technology (3)
		Computer Application Technology (1)
	Urban and Rural Planning (4)	Urban and Rural Planning (2)
		Urban Planning and Design (2)
	Surveying and Mapping (3)	N.A.[f]
	Civil Engineering (2)	
	Control Science and Engineering (2)	
	Landscape Architecture (1); Mineral Engineering (1); Forestry Engineering (1)	
History (42)	Chinese History (37)[g]	Modern Chinese History (4)
		Historical Geography (3)[h]
		Ancient Chinese History (3)
		Special History (17)
	World History (3)	N.A.[f]
	Archaeology (2)	Archaeology and Museology (1)
		Theory and History of Historiography (1)
Law (38)	Sociology (18)	Anthropology (11)
		Sociology (2)[i]
		Ethnic Sociology (2)
		Demography (2)
		Folkloristics (1)
	Ethnology (14)[e]	Chinese Ethnic Economy (9)
		Public Administration in Ethnic Areas (1)
		History of Chinese Ethnic Minorities (1)
	Law (4)[j]	Legal Theory (1)
		International Law (1)
		Environmental and Resource Protection Law (1)
		Economic Law (1)
	Political Science (1)	History of the Communist Party of China (1)
	Marxist Theory (1)	Ideological and Political Education (1)

(Continued)

Disciplinary cluster	First-level disciplines (no. of dissertations)	Second-level disciplines (no. of dissertations)
Agriculture (22)	Forestry (18)	Silviculture (6)
		Nature Reserve Science (4)
		Ornamental Plants and Horticulture (2)
		Soil and Water Conservation and Desertification Control (2)
		Forest Management (2)
		Forest Recreation and Park Management (1)
		Protection and Utilization of Wild Animals and Plants (1)
	Agricultural Resources and Environment (2)	Pedology (2)
	Animal Husbandry (1)	Pratacultural Science (1)
	Horticulture (1)	Tea Science (1)
Literature (19)	Chinese Language and Literature (9)[d]	Modern and Contemporary Chinese Literature (3)
		Ancient Chinese Literature (3)
		Linguistics and Applied Linguistics (1)
		Art and Literature (1)
	Foreign Language and Literature (7)	English Language and Literature (4)
		Comparative Literature and Cross Cultural Studies (1)
		German Language and Literature (1)
		Foreign Linguistics and Applied Linguistics (1)
	Journalism and Communication (3)	Communication (2)
		Journalism (1)
Education (13)	Physical Education (12)	Sports Humanities and Sociology (9)
		Physical Education Training (3)
	Psychology (1)	Basic Psychology (1)
Arts (8)	Art Theory (3)	Art Theory (3)
	Design (5)[c]	Design Art (3)
Medicine (4)	Traditional Chinese Medicine (4)[d]	Health Preservation of Traditional Chinese Medicine (2)
		Basic Theories of Traditional Chinese Medicine (1)
Philosophy (4)	Philosophy (4)	Ethics (2)
		Cultural Philosophy (1)
		Study of Religion (1)

a Since 2011, the Ministry of Education of China has allowed higher education institutions to autonomously set up second-level disciplines outside the existing Discipline Catalogue of Degree Granting (*Catalogue*, hereafter). In addition, there has been a trend for higher education institutions to recruit and cultivate graduate students at the first-level disciplinary level. Therefore, in this case, 12 PhD dissertations were submitted in fulfillment of the requirements of a first-level discipline (Business Administration) without indicating specific second-level disciplinary affiliation.

b In the *Catalogue*, Management Science and Engineering has no subordinating second-level disciplines. However, some higher education institutions create their own second-level disciplines, like in this case, the Industrial Engineering and Management Engineering.

c Two dissertations did not indicate their specific second-level disciplinary affiliation.

d One dissertation did not indicate its specific second-level disciplinary affiliation.

e Three dissertations did not indicate their specific second-level disciplinary affiliation.

f No specific second-level disciplinary affiliations were disclosed.

g Ten dissertations did not indicate their specific second-level disciplinary affiliation.

h A PhD candidate submitting a historical geography dissertation could be award a doctorate either in history or in geography.

i Sociology is a second-level discipline under first-level discipline *Sociology*.

j Law is also a first-level discipline.

Table 11.5 The major disciplinary subcategories of leisure-related doctoral dissertations in China

Disciplinary cluster	First-level disciplines (no. of dissertations)	Second-level disciplines (no. of dissertations)
Management (21)	Business Administration (10)	Tourism Management (7) Economics and Management of Technology (2) Enterprise Management (1)
	Management Science and Engineering (1)	Management Science and Engineering (1)
	Economic Management of Agriculture and Forestry (9)	Economic Management of Agriculture (5) Economic Management of Forestry (2) Economic Management of Agriculture Fishery (1) Operation and Management of Agricultural Enterprises (1)
	Engineering Management (1)	Resource Management Engineering (1)
Education (17)	Physical Education (16)	Sports Humanities and Sociology (7) Physical Education Training (6) Ethnic Traditional Sports (2) Kinesiology (1)
	Education (1)	Curriculum and Teaching Theory (1)
Philosophy (15)	Philosophy (15)	Leisure Studies (8) Aesthetics (4) Marxist Philosophy (2) Ethics (1)
Law (12)	Marxist Theory (7)	Basic principles of Marxism (3) Marxism in China (2) Ideological and Political Education (2)
	Ethnology (3)	Chinese Ethnic Economy (2) Chinese Ethnic Arts (1)
	Sociology (2)	Sociology (1); Anthropology (1)
Agriculture (10)	Forestry (4)	Silviculture (2) Ornamental Horticulture (1); Ornamental Plants and Horticulture (1)
	Crop Science (4)[a]	Agricultural Multifunctional Industry (1); Agricultural Science and Technology Service and Management (1); Agricultural Popularization (1)
	Agricultural Resources and Environment (1)	Plant Nutrition (1)
	Environmental Science and Engineering (1)	Horticultural Environmental Engineering (1)
Science (9)	Geography (4)	Human Geography (3) Physical Geography (1)
	Ecology (4)[b]	Agricultural Regional Development and Planning (1)
	Environmental Science and Engineering (1)[c]	Environmental Science (1)
Economics (5)	Applied Economics (4)[a]	Industrial Economics (1); National Economics (1); Regional Economics (1)
	Theoretical Economics (1)	Political Economics (1)

(Continued)

Disciplinary cluster	First-level disciplines (no. of dissertations)	Second-level disciplines (no. of dissertations)
Engineering (4)	Landscape Architecture (2)	N.A.[d]
	Urban and Rural Planning (1)	Urban Planning and Design (1)
	Textile Science and Engineering (1)	Fashion Design and Engineering (1)
History (3)	Chinese History (3)[a]	Special History (2)
Literature (3)	Chinese Language and Literature (2)	Ancient Chinese Literature (1); Modern and Contemporary Chinese Literature (1)
	Journalism and Communication (1)	Communication (1)

a One dissertation did not indicate its specific second-level disciplinary affiliation.
b Three dissertations did not indicate their specific second-level disciplinary affiliation.
c A PhD candidate submitting an environmental science and engineering dissertation could be awarded a doctorate either in science, engineering, or in agriculture.
d No specific second-level disciplinary affiliations were disclosed.

Table 11.6 The major disciplinary subcategories of hotel-related doctoral dissertations in China

Disciplinary cluster	First-level disciplines (no. of dissertations)	Second-level disciplines (no. of dissertations)
Management (40)	Business Administration (34)[a]	Tourism Management (24) Enterprise Management (5) Economics and Management of Technology (1)
	Management Science and Engineering (6)	Management Science and Engineering (5) Management Science (1)
Science (2)	Geography (1)	Human Geography (1)
	Ecology (1)	N.A.[b]
Economics (1)	Applied Economics (1)	International Marketing (1)
Engineering (1)	Forestry Engineering (1)	Wood Science and Technology (1)
History (1)	Chinese History (1)	Modern Chinese History (1)
Law (1)	Sociology (1)	Sociology (1)

a Four dissertations did not indicate their specific second-level disciplinary affiliation.
b No specific second-level disciplinary affiliations were disclosed.

(24) was the dominant second-level discipline, followed by Enterprise Management (5) and Economics and Management of Technology (1). Other disciplinary clusters such as *Science* (2), *Economics* (1), *Engineering* (1), *History* (1), and *Law* (1) have sporadically contributed to the doctoral studies on hotel issues.

Differences between research institutions

With respect to degree-granting institutions, as shown in Table 11.7, Sichuan University ranks No. 1 in the total number of doctoral dissertations pertaining

Table 11.7 Ranking of higher institutions in China by number of doctoral dissertations in the fields of tourism, leisure, and hotel

Higher institution	Tourism	Leisure	Hotel	Total
Sichuan University*	113	0	0	113
Sun Yat-sen University*	63	2	6	71
Xiamen University[a],*	41	3	3	46[a]
Shaanxi Normal University**	39	0	1	40
Northwest University**	28	0	1	29
Zhejiang University[a],*	13	13	2	27[a]
Fudan University*	23	1	2	26
Northeast Normal University**	25	1	0	26
Yunnan University*	24	1	0	25
East China Normal University[a],*	23	2	1	25[a]
Peking University*	23	2	0	25
Dongbei University of Finance and Economics	20	2	1	23
Wuhan University*	22	1	0	23
Beijing Forestry University**	16	5	0	21
China University of Geosciences[b],**	14	3	0	17[b]
Dalian University of Technology*	13	1	3	17
Jilin University*	13	3	1	17
Huaqiao University	9	0	8	17
Minzu University of China*	13	2	0	15
Central South University of Forestry and Technology[a]	12	2	1	14[a]
Ocean University of China*	7	7	0	14
Nanjing Normal University**	12	1	0	13
Tianjin University*	11	1	1	13
Nanjing University*	12	0	0	12
Lanzhou University*	11	0	0	11
Beijing Jiaotong University**	10	0	0	10
Central China Normal University**	10	0	0	10
Liaoning Normal University	9	1	0	10
Nankai University*	9	0	1	10
South China University of Technology*	7	1	2	10
Beijing Sport University**	4	6	0	10
Anhui Normal University	9	0	0	9
Southwest Jiaotong University**	9	0	0	9
Shandong University[a]*	7	3	0	9[a]
Chengdu University of Technology**	8	0	0	8
Hunan Normal University**	6	1	1	8
Northeast Forestry University**	5	1	1	7
Henan University**	6	0	0	6
Hunan University*	6	0	0	6
Yanshan University	6	0	0	6
Chongqing University*	5	0	1	6
University of International Business and Economics**	5	0	1	6
Wuhan University of Technology**	5	0	1	6
Fujian Normal University	4	2	0	6
Nanjing Forestry University[c],**	5	2	0	5[c]
Zhejiang Gongshang University	5	0	0	5
Zhongnan University of Economics and Law**	5	0	0	5

(Continued)

Higher institution	Tourism	Leisure	Hotel	Total
Harbin Institute of Technology*	4	1	0	5
University of Science and Technology of China*	3	0	2	5
Fujian Agriculture and Forestry University	1	4	0	5

a   One doctoral dissertation which has both leisure and tourism themes was accounted only once in the total number to avoid double counting.
b   The number of China University of Geosciences includes that of CUG (Wuhan) and CUG (Beijing).
c   Two doctoral dissertations that have both leisure and tourism themes were accounted only once in the total number to avoid double counting.
*   China's top universities as identified in the Chinese government's 'Top University Plan'.
** China's top universities as identified in the Chinese government's 'Top Discipline Plan'; see https://en.wikipedia.org/wiki/Double_First_Class_University_Plan, accessed 17 February 2020.

to tourism in the broader sense (113 dissertations, including 'tourism', 'leisure', and 'hotel'), all of which are concerned with 'tourism' topics. A closer look suggests that there are at least four major second-level disciplines that contribute to doctoral studies in Sichuan University, namely *Tourism Management* (55 dissertations; 48.7%), *Enterprise Management* (13 dissertations; 11.5%), *Chinese History* (26 dissertations; 23%), and *Management Science and Engineering* (eight dissertations; 7.1%), among others.

Sun Yat-sen University follows with 63 dissertations in 'tourism', two in 'leisure', and six in 'hotel', among which *Tourism Management* is also the leading contributive discipline. Specifically, in the 'tourism' category, 46 out of 63 (73%) dissertations were submitted in *Tourism Management*, followed by *Human Geography* (8), *Anthropology* (5), *English Language and Literature* (1), *Ethics* (1), *Finance* (1), and *Physical Geography* (1). In the hotel category, *Tourism Management* is still the dominant contributing discipline, with five out of the six dissertations submitted in this discipline, while the other one is in *Enterprise Management*. In addition, the two leisure-focused dissertations are conducted within the disciplinary backgrounds of *Human Geography* and *Ideological and Political Education*, respectively.

Xiamen University ranks third in terms of the total number of doctoral dissertations pertaining to tourism in the broader sense. Specifically, among the 46 dissertations, 40 are with an exclusive focus on tourism, two on leisure, and three on hotel. In addition, one dissertation covers both tourism and leisure. With respect to disciplinary backgrounds, 33 dissertations were submitted within *Tourism Management* (71.7%), six were within *Anthropology*, and two were within *Sociology*, while the remaining were within *Communications, English language and Literature, Ethnology*, and *Public Administration*.

Shaanxi Normal University and Northwest University, located in Xi'an, are both well-established tourism research institutions in China. Both universities have graduated a great number of tourism PhD students. In the past decade (2010–2019), Shaanxi Normal University recorded 40 PhD completions:

39 dissertations were focused on tourism and one on hotel. Furthermore, 36 (90%) of those dissertations were completed within *Tourism Management* and the remaining four were within *Human Geography, Physical Geography, Natural Disaster Science*, and *Historical Geography*. Similarly, among the 29 doctoral dissertations that Northwest University completed during this period, 28 of them were with a tourism focus and the remaining one with a hotel focus. Regarding disciplinary backgrounds, 19 of the dissertations were submitted under *Tourism Management, six under Human Geography*, and the rest under *Physical Geography, Western Economics, Archaeology and Museology*, and *Economics of Population, Resources and Environment*, respectively.

Of particular note is Zhejiang University, which completed 27 doctoral dissertations during 2010–2019. Among them, 12 focused exclusively on tourism issues, two on hotel issues, one with a dual focus on both tourism and leisure, and another 12 focused exclusively on leisure issues, making Zhejiang University the No. 1 institution in producing leisure-themed doctoral dissertations. This has mainly been enabled by the fact that Zhejiang University established its own second-level discipline *Leisure Studies* under *Philosophy*, within which eight leisure-themed doctoral dissertations were produced. Another three leisure-themed doctoral dissertations were submitted within *Esthetics*, which is also a second-level discipline under *Philosophy*.

In addition, some top research universities, such as Fudan University (26), East China Normal University (25), Peking University (25), and Wuhan University (23), together with those less research-oriented universities such as Northeast Normal University (26), Yunnan University (25), Dongbei University of Finance and Economics (23), Beijing Forestry University (21), and Huaqiao University (17), have all greatly contributed to the production of China's doctoral dissertations pertaining to the tourism, leisure, and hotel fields. Such doctoral research contributions are a significant part of China's tourism, leisure, and hotel studies, with their own disciplinary backgrounds and characteristics. For instance, most of Beijing Forestry University's doctoral dissertations pertaining to the tourism, leisure, and hotel fields were completed in two first-level disciplines, that is, Forestry (10) and Economic Management of Agriculture and Forestry (6). As a comprehensive university and with Business Administration (including Tourism Management) as one of its strength disciplines, Huaqiao University's 17 doctoral dissertations (i.e., in tourism and hotel) were all completed within the discipline of Business Administration.

Tourism research institutions

Among the four major types of tourism research institutions, higher educational institutes (i.e., universities and colleges) and specialized research institutions, instead of corporations and governmental agencies, have been increasingly playing a dominant role, especially in terms of publications, research talent cultivation, and grants (e.g., Chen and Huang, 2020; Huang and Chen, 2016a; Huang et al., 2019). As shown in Table 11.8, while most of the leading tourism research

Table 11.8 Leading tourism research institutions in China by publications in Chinese

Institution	Rank by H index (score)	Rank by journal articles (score)	Rank by research grants (number)
Anhui Normal University	1 (69)	6 (331.35)	4 (21)
Sun Yat-sen University*	2 (67)	2 (580.68)	1 (42)
Shaanxi Normal University**	3 (62)	1 (711.27)	5(20)
Nanjing University*	4 (59)	8 (311.88)	12 (13)
Chinese Academy of Sciences	5 (57)	3 (490.98)	N.A.
Nanjing Normal University**	6 (52)	5 (354.44)	12 (13)
East China Normal University*	7 (50)	7 (327.49)	31 (7)
Peking University*	8 (48)	10 (269.52)	N.A.
Beijing Union University	9 (41)	9 (295.64)	9 (16)
Beijing International Studies University	10 (40)	21 (198.36)	9 (16)
Zhejiang University*	10 (40)	17 (209.61)	25 (8)
Shanghai Normal University	10 (40)	15 (211.27)	17 (9)
Henan University**	13 (39)	25 (171.18)	25 (8)
Northwest University**	13 (39)	24 (185.38)	25 (8)
Yunnan University*	15 (37)	12 (249.88)	5 (20)
Jinan University**	16 (36)	11 (255.46)	17 (9)
Zhejiang Gongshang University	16 (36)	27 (166.49)	17 (9)
Fudan University*	18 (35)	44 (117.17)	25 (8)
Sichuan University*	18 (35)	4 (390.28)	14 (11)
Nankai University*	18 (35)	20 (198.91)	7 (19)
Jishou University	18 (35)	13 (237.03)	11 (15)
Central South University of Forestry and Technology	22 (34)	16 (210.02)	N.A.
Huaqiao University	23 (33)	14 (226.44)	8 (17)
Chinese Academy of Social Sciences	23 (33)	23 (191.88)	N.A.
Hubei University	23 (33)	41 (124.46)	34 (6)
Xiamen University*	26 (32)	22 (197.64)	34 (6)
Dongbei University of Finance and Economics	27 (31)	36 (128.20)	18 (9)
Northwest Normal University	28 (30)	26 (170.47)	N.A.
Shandong University*	29 (29)	32 (142.51)	34 (6)
China Tourism Academy	36 (27)	28 (153.14)	2 (32)

Data source: Zhang et al. (2019) for H index and journal articles (2003–2018); Chen and Huang (2020) for research grants (2010–2018); doctoral dissertations are compiled by the authors (2010–2019).
Notes: * China's top universities as identified in the Chinese government's 'Top University Plan'; ** China's top universities as identified in the Chinese government's 'Top Discipline Plan'; see https://en.wikipedia.org/wiki/Double_First_Class_University_Plan, accessed 17 February 2020.

institutions are universities, the Chinese Academy of Sciences, the Chinese Academy of Social Sciences, and the China Tourism Academy have all played their critical roles in receiving research grants and publishing journal articles.

As a leading research university in China, Sun Yat-sen University has been playing a comprehensive leading role in China's tourism research in four aspects (i.e., research influence, article numbers, research funds, and doctoral graduates). In addition, Shaanxi Normal University has also been an all-around leader in China's tourism research; especially, it occupies a leading position in terms of the number (frequency) of articles published in Chinese journals. Among the listed 31 institutions, 11 are top universities as identified in the Chinese government's 'Top University Plan' and another five are listed in the 'Top Discipline Plan', which indicates the relatively high status of the tourism discipline in China.

Notably, mainland Chinese universities have been playing an increasingly important role in the world's tourism research, which could be partly reflected by their ranks in internationally recognized rankings, for instance, the Global Ranking of Academic Subjects by ShanghaiRanking. In 2020, Sun Yat-sen University was ranked No. 4 globally in this ranking, a big progress compared to its positions in previous years: No. 8 in 2019, No. 10 in 2018, and No. 17 in 2017. In addition, other top universities in the Chinese mainland have also won more recognition. Specifically, in 2020, Zhejiang University and Xiamen University were ranked in the subject area of Tourism and Hospitality Management in ShanghaiRanking as No. 28 (No. 51–75 in the 2019 ranking) and No. 36, respectively. Other universities in mainland China that are included in the 2020 ShanghaiRanking list in the subject of Tourism and Hospitality Management include Jinan University in Guangzhou (No. 51–75), Nanjing University (No. 76–100), Beijing Union University (No. 101–150), Harbin Institute of Technology (No. 101–150), Peking University (No. 101–150), and Shanghai University of Sport (No. 201–300).

It is important to note that tourism research in China has been driven not only by those top research universities, such as those listed in the government's 'Top University Plan' and 'Top Discipline Plan', but also by those so-called 'local universities' and those universities that have not been preferentially supported by the central government (see Table 11.8), such as Anhui Normal University, Beijing International Studies University, Beijing Union University, Shanghai Normal University, Zhejiang Gongshang University, Jishou University, Central South University of Forestry and Technology, Huaqiao University, Hubei University, and Dongbei University of Finance and Economics, among many others. These universities, with their disciplinary characteristics and local strengths, have produced a great number of high-quality research outputs, conducted numerous fruitful consulting projects for local tourism authorities and enterprises, and helped cultivate generations of China's tourism research talents.

Tourism researchers

China may have the largest number of tourism researchers. According to Zhang et al. (2019), there were a total of 21,854 authors who have published at least one tourism-related paper in a Chinese academic journal between 2003 and 2018. From 2003 to 2016, this figure was 18,773 (Zhang et al., 2019). This simply means that at the end of 2018, the number of Chinese tourism researchers (including students) had an increase of 16.4% from the end of 2016.

Most influential tourism researchers

In terms of influential tourism researchers, the past two years have also seen big changes. For instance, during 2003–2018 (Zhang et al., 2019), among the top 100 tourism researchers in terms of publication frequency, 11 were newly included in the list, in comparison to 2003–2016 (Zhang, Wang, Zhang, Han, and Zhang, 2017). Furthermore, among the 11 researchers, two were (i.e., Cao Xinxiang and Zhang Hongmei) ranked among the top 50 most influential tourism researchers in China by H index (Table 11.9).

As shown in Table 11.9, three researchers (i.e., Bao Jigang, Lu Lin, and Wu Bihu) from Sun Yat-sen University, Anhui Normal University, and Peking University, respectively, who all hold a geography discipline background, were continuously identified as China's most influential tourism researchers in terms of the impacts (i.e., H index) of their research outputs published in Chinese journals. Among the top 50 most influential tourism researchers by H index (Zhang et al., 2019), there are three (6%) younger researchers who were born in the 1980s, 17 (34%) in the 1970s, 23 (46%) in the 1960s, 5 (10%) in the 1950s, and 2 (4%) in the 1940s. Compared with the generational distribution of the 87 most influential tourism researchers (1980s: 0; 1970s: 28.7%; 1960s: 43.7%; 1950s: 17.2%; 1940s: 3.4%) as identified in Chen and Huang (2020), influential tourism researchers in China have slightly shifted to be the much younger generations.

Categories of tourism researchers

Due to differences in academic training, disciplinary backgrounds, and career development pursuits, tourism researchers in China are different and can thus be classified into different categories (Bao et al., 2019; Chen, 2019). According to Chen (2019), there are at least three categories of tourism researchers in China. The first category includes those researchers whose backgrounds are rooted in traditional disciplines, such as geography, sociology, anthropology, and management, and for whom tourism is just one of their many research areas of interest. To them, tourism is similar to what Pearce (2011) referred to as 'a serendipitous finding'. A typical researcher is Professor Wang Ning from Sun Yat-sen University, whose disciplinary background is sociology. His areas of interest include consumption, institution, and tourism, all with a sociological perspective. Researchers in this category can be found across different generations of tourism researchers in China (Bao et al., 2019). In the *pioneer generation* (Weng and Bao, 2017), Professor Chen Chuankang and Professor Guo Laixi may be classified into this category; in the *first generation*, Professor Zhu Hong and Professor Zhou Shangyi could fit into this category; in the *middle generation* and *new generation*, Professor Su Xiaobo (geography), Dr. Qian Junxi (geography), Dr. Weng Shixiu (geography), Professor Cai Xiaomei (geography), and Dr. Chen Zengxiang (consumer behavior) are typical tourism researchers in this category. It should be noted that for tourism researchers in this category, one biggest advantage is that they have a specific disciplinary background and may therefore be specifically equipped within a discipline in both theoretical and methodological terms. The other advantage is that they have a risk dispersion mechanism. That is, they 'don't put eggs in the same basket' in their research.

Table 11.9 The top 50 most influential tourism researchers in China by publications in Chinese

Researcher name	Rank in Zhang et al. (2019)	Rank in Zhang et al. (2017)	Institution	Born in
Bao, Jigang	1	1	Sun Yat-sen University*	1960s
Lu, Lin	2	2	Anhui Normal University	1960s
Wu, Bihu	3	3	Peking University*	1960s
Wang, Degen	4	8	Soochow University**	1970s
Bian, Xianhong	5	5	Zhejiang Gongshang University	1970s
Ma, Xiaolong	6	7	Nankai University*	1970s
Ma, Yaofeng	7	6	Shaanxi Normal University**	1940s
Zhang, Jie	8	4	Nanjing University*	1960s
Sun, Gennian	9	9	Shaanxi Normal University**	1960s
Wang, Zhaofeng	10	10	Hunan Normal University**	1960s
Bai, Kai	11	12	Shaanxi Normal University**	1970s
Huang, Zhenfang	12	11	Nanjing Normal University**	1960s
Sun, Jiuxia	13	19	Sun Yat-sen University*	1960s
Xu, Honggang	14	13	Sun Yat-sen University*	1960s
Guo, Lufang	15	14	Zhejiang Gongshang University	1960s
Zhang, Jinhe	16	21	Nanjing University*	1970s
Yang, Xinjun	17	15	Northwest University**	1970s
Zhang, Hongmei	18	17	Anhui Normal University	1960s
Zhong, Linsheng	19	30	Chinese Academy of Sciences	1970s
Ma, Yong	20	25	Hubei University	1950s
Ma, Lijun	21	23	Xiangtan University	1980s
Zheng, Xiangmin	22	18	Huaqiao University	1950s
Cao, Xinxiang	23	N.A.	Henan University**	1970s
Zhang, Guanghai	24	33	Ocean University of China*	1960s
Yang, Zhenzhi	25	22	Sichuan University*	1960s
Zhu, Hong	26	16	South China Normal University**	1960s
Feng, Xuegang	27	20	East China Normal University*	1960s
Li, Junyi	28	37	Shanghai Normal University	1970s
Tao, Wei	29	29	South China Normal University**	1970s
Zhang, Lingyun	30	24	Beijing Union University/Beijing International Studies University	1960s
Xie, Yanjun	31	35	Dongbei University of Finance and Economics/Hainan University	1960s
Su, Qin	32	52	Anhui Normal University	1960s
Lu, Song	33	28	Shanghai Normal University	1970s
Liu, Jiaming	34	46	Chinese Academy of Sciences	1960s
Zhang, Chaozhi	35	34	Sun Yat-sen University*	1970s
Liang, Mingzhu	36	27	Jinan University**	1950s
Ma, Bo	37	N.A.	Qingdao University	1960s
Zhao, Lei	38	31	Zhejiang University of Technology	1980s
Jin, Cheng	39	61	Nanjing Normal University**	1980s
Chen, Tian	40	43	Chinese Academy of Sciences	1950s
Li, Donghe	41	39	Anhui Normal University*	1970s
Dong, Guanzhi	42	26	Jinan University**	1960s
Huang, Fucai	43	45	Xiamen University*	1940s
Shi, Chunyun	44	40	Jiangsu Normal University	1970s
Yang, Yong	45	41	East China Normal University*	1970s
Xi, Jianchao	46	47	Chinese Academy of Sciences	1970s
He, Jianmin	47	44	Shanghai University of Finance and Economics**	1950s
Yang, Xiaozhong	48	55	Anhui Normal University	1960s
Wang, Qun	49	60	Anhui Normal University	1970s
Lu, Yuqi	50	54	Nanjing Normal University**	1960s

Note: Ranks in Zhang et al. (2019) and Zhang et al. (2017) were both based on H index. * China's top universities as identified in the Chinese government's 'Top University Plan'; ** China's top universities as identified in the Chinese government's 'Top Discipline Plan'; see https://en.wikipedia.org/wiki/Double_First_Class_University_Plan, accessed 17 February 2020.

A second category refers to those researchers who also study tourism from a traditional discipline (e.g., geography, economics, anthropology, and management) but for whom tourism may be the only or at least dominant research area of expertise. Similarly, in the *first generation* (Bao et al., 2019; Weng and Bao, 2017), Professor Bao Jigang, Professor Lu Lin, Professor Wu Bihu, and Professor Zhang Jie are all representative researchers from geography; in the *middle generation*, Professor Sun Jiuxia and Zhao Hongmei are representatives from anthropology. In the *new generation*, Zhang Honglei (geography), Yang Yang (geography and economics), and Yang Yong (economics) fit into this category. While this category of tourism researchers is specifically equipped within a discipline in both theoretical and methodological terms, they may face a certain degree of risk: exclusively conducting tourism research from a single discipline. One trend that should be noted is that it seems an increasing number of younger tourism researchers are 'getting away' from tourism or at least treating tourism as just one of their diversifying research areas.

The third category is what has been widely referred to as *Generation T* (tourism) researchers. Understandably, most of them are educated and academically trained in the discipline of *Tourism Management*. Therefore, the majority of their academic training is around tourism research and may be around a specific source discipline, such as geography, economics, and anthropology (depending on the backgrounds of their supervisors' and/or disciplinary backgrounds). Different from the above two categories, *Generation T* researchers may be better equipped with understandings of both the tourism literature and tourism industry. It is however apparent that this group of researchers is getting gradually divided. An increasing number of doctoral dissertations are conducting research from disciplines (e.g., geography, economics, sociology, and anthropology) other than management and thus 'getting away' from management, although the degree-granting discipline is *Tourism Management*. Under this circumstance, *Generation T* researcher may increasingly face a disciplinary identity crisis. Meanwhile, dilemma may also undermine the disciplinary status of *Tourism Management* within *Business Administration* and *Management*, which is already in danger to some degree.

Conclusion

Using doctoral dissertations as a data source, this chapter identified the major disciplinary backgrounds pertaining to tourism research in China. *Management* (disciplinary cluster), *Business Administration* (first-level discipline), and *Tourism Management* (second-level discipline) are, in most cases, dominant disciplines in tourism-related (i.e., tourism, leisure, and hotel) doctoral dissertations in their respective disciplinary level. Universities may have their disciplinary characteristic in cultivating China's next generation of tourism researchers (i.e., doctorate holders) due to historical and local circumstances. Meanwhile, in terms of doctoral dissertations, publications in both Chinese-language and English-language journals as well as research impacts, leading research universities in China such as Sun Yat-sen University, Nanjing University, Zhejiang University, and Xiamen University have been playing a leading role in China's tourism research and have been playing an increasingly important role in the world's tourism research, partly reflected by their climbing positions in internationally recognized rankings (e.g., the ShanghaiRanking's Global Ranking of Academic Subjects). It is important to note that tourism research in China has been greatly driven by those so-called 'local universities' and those universities that have not been preferentially supported by the central government. Influential tourism researchers in China have included younger generation members. It is believed that tourism researchers of different categories with varying backgrounds of academic training, disciplinary perspectives, and career development pursuits have been (re) shaping and will (re)shape tourism research in China.

References

Afifi, G M H 2013, 'A survey of doctoral theses accepted by universities in the United Kingdom and Ireland for studies related to tourism, 2000–2009', *Journal of Hospitality & Tourism Education*, vol. 25, no. 1, pp. 29–39.

Bao, J 2002, 'Tourism geography as the subject of doctoral dissertations in China, 1989–2000', *Tourism Geographies*, vol. 4, no. 2, pp. 48–152.

Bao, J & Ma, L J C 2011. 'Tourism geography in China, 1978–2008: Whence, what and whither?' *Progress in Human Geography*, vol. 35, no. 1, pp. 3–20.

Bao, J, Chen, G & Jin, X 2018, 'China tourism research: A review of publications from four top international journals', *Journal of China Tourism Research*, vol. 14, no. 5, pp. 1–19.

Bao, J, Chen, G & Ma, L 2014, 'Tourism research in China: Insights from insiders', *Annals of Tourism Research*, vol. 45, no. 1, pp. 167–181.

Bao, J, Huang, S & Chen, G 2019, 'Forty years of China tourism research: Reflections and prospects', *Journal of China Tourism Research*, vol. 15, no. 3, pp. 283–294.

Bao, J & Lai, K 2016, 'On the disciplinary connotations of Tourism Management and the needs for its upgrading', *Tourism Tribune*, vol. 31, no. 10, pp. 14–16.

Chen, D 2004, 'The disciplinary development of tourism from the perspective of tourism-related doctoral dissertations', *Tourism Tribune*, vol. 19, no. 6, pp. 9–14.

Chen, G 2019, 'Tourism disciplinary development in China: Disciplinary arrangement and researcher groups', *Tourism Tribune*, vol. 34, no. 12, pp. 7–9.

Chen, G & Huang, S 2020, 'Tourism research in China' in S Huang & G Chen, (eds), *Handbook of Tourism and China*, pp. 356–381. Edward Elgar Publishing, Cheltenham.

Huang, S 2011, 'Tourism as the subject of China's doctoral dissertations', *Annals of Tourism Research*, vol. 38, no. 1, pp. 316–319.

Huang, S 2019, 'Tourism higher education and disciplinary development in Australia', *Tourism Tribune*, vol. 34, no. 11, pp. 8–11.

Huang, S & Chen, G 2016a, *Tourism research in China: Themes and issues*, Channel View Publications, Bristol.

Huang, S & Chen, G 2016b, 'Current state of tourism research in China', *Tourism Management Perspective*, vol. 20, pp. 10–18.

Huang, S & Chen, G (eds) 2020, *Handbook of Tourism and China*, Edward Elgar Publishing, Cheltenham.

Huang, S, Chen, G, Luo, X & Bao, J 2019, 'Evolution of tourism research in China after the millennium: Changes in research themes, methods and researchers', *Journal of China Tourism Research*, vol. 15, no. 3, pp. 420–434.

Lu, X, Chen, X & Ma, B 2016, 'Tourism academic research streams based on doctoral dissertations', *Research in Higher Financial and Economics Education*, vol. 19, no. 1, pp. 85–94.

Ma, B 2016, 'On the approach to upgrading the disciplinary status of tourism in China', *Tourism Tribune*, vol. 31, no. 10, pp. 25–32.

Ministry of Culture and Tourism 2020, PRC Ministry of Culture and Tourism 2019 yearly report on culture and tourism, 22 June 2020. Available from: http://www.gov.cn/shuju/2020-06/22/content_5520984.htm [31 July 2020].

Pearce, P L (Ed) 2011, *The study of tourism: Foundations from psychology*, Emerald, Bingley.

Shen, S & Philippe, V 2019, 'Tourism talent development at master and doctoral levels and disciplinary development in France', *Tourism Tribune*, vol. 34, no. 11, pp. 6–8.

Tang, S 2013, 'An analysis and new research directions of doctoral dissertations on tourism in China in the recent decade', *Tourism Tribune*, vol. 28, no. 3, pp. 106–113.

Weiler, B, Moyle, B & McLennan, C 2012, 'Disciplines that influence tourism doctoral research: The United States, Canada, Australia and New Zealand', *Annals of Tourism Research*, vol. 39, no. 3, pp. 1425–1445.

Weng, S & Bao, J 2017, 'The cross-generational differences and transformation of the academic practices of tourism geography in China', *Geographical Research*, vol. 36, no. 5, pp. 824–836.

Wu, Z, Wang, Z & Lin, W 2019, 'Tourism-related disciplinary development history, situation, and challenges: Also discussing faculty and research capacity', *Tourism Tribune*, vol. 34, no. 11, pp. 13–16.

Xiao, H & Smith, S 2006, 'The making of tourism research: Insights from a social sciences journal', *Annals of Tourism Research*, vol. 33, no. 2, pp. 490–507.

Yang, W & Mao, Z 2019, 'Tourism-related disciplinary development in the US and its implications to China', *Tourism Tribune*, vol. 34, no. 11, pp. 4–6.

Yang, Y & Xu, X 2017, 'Tourism scholar's academic influence evaluation and discipline development: Based on statistical analysis of tourism academic articles in CNKI database', *Tourism Tribune*, vol. 32, no. 9, pp. 103–115.

Zhang, X & Chen, X 2019, 'Tourism-related disciplinary development and research in the UK', *Tourism Tribune*, vol. 34, no. 11, pp. 1–3.

Zhang, L, Han, L, Zhang, Y, Zhou, X & Zhang, S 2019, 'Evaluating the research performances of tourism academic communities of China from 2003 to 2018', *Tourism Tribune*, vol. 34, no. 12, pp. 120–136.

Zhang, L, Wang, C, Zhang, D, Han, L & Zhang, Y 2017, 'Evaluating the research performances of tourism academic communities of China from 2003 to 2016', *Tourism Tribune*, vol. 32, no. 12, pp. 117–127.

Zhang, Y & Xu, N 2019, 'Tourism-related disciplinary development and specific program arrangement', *Tourism Tribune*, vol. 34, no. 11, pp. 11–13.

12 Critical issues, challenges, and future prospects

Songshan (Sam) Huang and Jigang Bao

Introduction

In the phase of this volume's planning and preparation, COVID-19 was still an unknown term. The world changed contrastingly in the period of preparing the chapters of this book, entering into the COVID-19 pandemic period. COVID-19 has now become such a significant factor changing the track of the world's development, let alone the development pathway of hospitality and tourism education in China. In this chapter, we re-evaluate a number of critical issues and challenges in association with hospitality and tourism development in China. Now that COVID-19 has substantially changed the landscape of international education worldwide, apparently hospitality and tourism education in China will not be immune to the impact of the COVID-19 pandemic. The pandemic appears to be a shock factor to forge the further development of China's hospitality and tourism education. Thus, we will also consider the impact of the COVID-19 pandemic in our discussions of the critical issues in this chapter. In a similar vein, the future of China's hospitality and tourism education may be an extension of the current reshuffled state due to the influence of the pandemic and the derived industry reformation and international geopolitical relations. Therefore, in our speculations of possible scenarios of China's hospitality and tourism education in the future, we also weighed in the influence of the COVID-19 pandemic.

Enrolment scales and structural change

As shown in Figure 1.4, Chapter 1, the student enrolment scale of China's hospitality and tourism education system, including both tertiary education and secondary vocational education levels, reached its peak in the years 2010 and 2011. After 2011, the number of student enrolment in hospitality and tourism programs kept declining. In terms of the number of institutions, while the number of higher education institutions kept growing, the secondary vocational schools began to decrease after 2012. In 2017, there were 1,694 higher learning institutes and 947 secondary vocational schools offering hospitality and tourism programs.

DOI: 10.4324/9781003004363-12

The declining student enrolments may be attributed to the demographic shift of the youth population in China. The first only-child generation in China has grown to be around in their 40s and China's population structure is shifting to see reducing shares of young people in the population pyramids. In 2010, the proportions of people in the age brackets of 15–19 and 20–24 in China's total population were 7.2% (3.8% male vs. 3.4% female) and 9.5% (4.9% male vs. 4.6% female), respectively, whilst in 2020, these proportions reduced to 5.8% (3.1% male vs. 2.7% female) and 6% (3.2% male vs. 2.8% female), respectively (Data source: www.populationpyramid.net/china). China's further massification of its higher education system may also be a reason. In the waves of recruitment expansion of Chinese universities, tourism management was among the many programs that saw a booming expansion of enrolment. The sudden large increase of student numbers in tourism programs may have created a ceiling effect. With more competing university majors and programs being added in the learning subject list, students may be diverted to other study subjects which may seem to be more promising in the evolving job markets. Before the COVID-19 pandemic, students also had the choice to study a hospitality and tourism major in foreign countries like Australia, UK, and US. These factors together may have partially contributed to the declining trend of China's hospitality and tourism education scale during 2011–2017.

Looking forward, we may see the declining trend continue in the coming years or decade. It is reasonable to speculate with border closures and travel restrictions, some students intending to study overseas may eventually choose to study in China, thus adding to the enrolments of tourism education in China. This effect may not be significant enough to reverse the dwindling trend of China's hospitality and tourism education.

The implications of such a trend on tourism education institutions and educators in China should not be underestimated. About a decade ago, some universities in China had already found it challenging to maintain student numbers in their hospitality and tourism programs. It wouldn't be surprising to see hospitality and tourism programs cut and discontinued in universities due to decreasing student enrolment numbers. Internationally, hospitality and tourism program closures and mergers due to lack of student demand are not new (Fidgeon, 2010). Time may have come to see such program closures and mergers happening more often in China's education system.

Of course, there may be a further issue with the structural distribution of students between the secondary- and tertiary-level institutions. If the tourism industry's further development favours vocational skills and the job market expects more vocational school graduates, there may still be a likely resurge of vocational tourism schools. The transition of China tourism industry development toward the vast rural areas may see the need of preparing more rural origin labour in the tourism sector. And secondary vocational tourism education may find a new space in the rural area with the transition and shift of the country's overall economic structure.

Curriculum

Curriculum is always the key part of the teaching and learning process and will directly determine the learning outcomes. Much of the discussion around low-quality textbooks in China's tourism education system is also related to the curriculum. Internationally, debates on tourism curriculum design and contents have been around how to balance or find a balanced structure over liberal arts and vocational needs in a neoliberal institutional environment (Ayikoru, Tribe, & Airey, 2009; Dredge, Benckendorff, Day, Gross, Walo, Weeks, & Whitelaw, 2012; Sheldon, Fesenmaier, & Tribe, 2011). Such a debate, however, has not been clearly witnessed among tourism educators in China. Curriculum design and development issues have been one of the concerns in the tourism education community (Yin & Meng, 2018; Zhang & Fan, 2005, Zins & Jang, 2019). And as shown in Chapter 9, there is pronounced difference between the curriculum contents in Chinese universities and that in Western universities.

Zins and Jang (2019) conducted a systemic review of the curriculum contents of hospitality and tourism programs in China. They found there was low transparency in the curriculum design and most programs examined lacked a clear profile and positioning of their programs. Commonly neglected contents in the curricula in the programs are personal and social skills such as problem solving, critical thinking, creativity, teamwork, and liberal reflection capacities. The authors' interviews with human resource managers of internationally branded hotels in China also revealed that personal qualities like emotional intelligence, an outgoing personality, and resilience are preferred from the employers.

While there is a general consensus among tourism educators in China that practice-based learning is important and should be reflected more in the program and curriculum design (Yin & Meng, 2018), there seems to be a lack of awareness of the shift from a vocational employability focus to a broader social ethics and values orientation in curriculum design in the tourism education community (Zins & Jang, 2019). In this regard, the Tourism Education Futures Initiative (TEFI) (Sheldon et al., 2011), advocating a shift of tourism education toward social responsibility, citizenship, and personal development, would be relevant to guide further discussions in and attention to curriculum reform among tourism educators in China.

Teachers

With about 40 years of development, the quality of the teaching staff in China's hospitality and tourism education system has been improved. However, the overall teaching workforce may vary significantly in their qualifications. While research-oriented and top tier universities may have most of teaching staff hold a doctoral degree, this is not the case with lower tier universities. According to a survey conducted by the China Tourism Education Association, 23.87% of the teaching staff in the survey tourism higher education institutions held a doctoral degree, 57.08% held a master's degree, while 19.05% did not have a postgraduate education (Yu & Zeng, 2016).

As elaborated by Xu and Zhang in Chapter 8, China's doctoral education in tourism and hospitality management has been developed quickly since the turn of the century. Currently, there are about 30 institutions offering PhD programs in tourism management, 13 institutions offering tourism PhD programs in the discipline of business administration, and another ten institutions offering tourism PhD programs in the discipline of human geography. As shown in Table 1.1 of Chapter 1, the number of enrolled doctoral students in China's tourism education reached 336 in 2017. The rapid expansion of tourism PhD education in China will provide more tourism PhD graduates to enter the workforce of teaching staff in tourism education. An increasing number of returned tourism PhD graduates from overseas institutions will further add to the tourism teaching workforce in China. With the dwindling enrolment scales in undergraduate and secondary vocational hospitality and tourism programs and the fast-growing tourism PhD enrolments in recent years, the whole tourism education workforce may be reshuffled with more tourism PhD graduates entering the field of tourism teaching. Accordingly, the percentage of teachers holding a PhD degree in the hospitality and tourism program may rise.

It is expected that most of the tourism PhD graduates, if they ever enter into the field of tourism teaching, would be a member of the Generation T. Generation T members in the tourism academic field are defined as those who obtained their first undergraduate degree, master's degree, and doctoral degree in the field of tourism. Different from the first- or second-generation tourism educators who have a parental discipline outside tourism, Generation T tourism teachers may have a knowledge structure centred (and thus limited) to tourism as a nascent knowledge field (Tribe, 2006). Two issues may persist to be challenges around the Generation T tourism teachers in the future. One, they may still lack the industry work experience after many years of formal education in the field of tourism, and thus may still find it difficult to address the vocational and employability needs from students in their teaching practice. Two, without a strong traditional discipline foundation in their education, they may not possess the required liberal arts spirit and critical thinking to prepare students for good citizenship attitude and ethics, as required by the needs of sustaining the industry in the future (Sheldon et al., 2011).

However, if the enrolment scales of tourism students keep declining, it is likely some hospitality and tourism programs in the higher education sector will have to be cut and removed. In some cases, demand for teachers will be restricted by the reduced scale of student enrolment. Teachers in these programs will have to find their way out, either going into a different but related field, for example, business administration, or moving into a different university to join another tourism department or school. To conclude, the future scenarios regarding tourism educators and teachers in China would be dynamic and largely determined by the forces of student demand and teacher supply as well as institutional change in China's higher and vocational education sectors.

Students

Irrespective of the possible declining trend of the scale of China's tourism education, students already remain to be the core stakeholder group of such an education system. There are some far-reaching issues with regards to tourism students in China. The foremost important issue is employability. On the one hand, employability will be determined by the industry's development trend. As long as there is a further development of the tourism industry in China, the industry demand for qualified workers will not disappear. Under the current context of the global COVID-19 pandemic, China has exhibited admiring capability of containing the COVID-19 infections and recovers its tourism industry amid the pandemic. Therefore, in the short term, the tourism industry would still need the labour supply, thus opening the avenue for the sustainable development of the tourism education system.

It has been observed that many students graduated from tourism programs do not choose to work in the tourism industry. One of the reasons is the low start salary level in the industry. Many bachelor degree graduates, especially those from key universities, are not willing to work in the hotel industry any longer after having hotel internship experience. Employment situation in the tourism industry has changed drastically. In the 1980s and the 1990s, tourism industry employment attracted a lot of young talents, as the salary level at that time was relatively high. In China's current development stage of industrialisation, urbanisation, and informatisation, the start salaries of many industries would be higher than that of the traditional tourism industry. As such, employment in the tourism industry appears less attractive to university graduates. Some excellent graduates from the tourism programs even find they can transfer their knowledge and skills to fields like marketing and pursue a better paid career in other service sectors.

On the other hand, employability will also be affected by the graduate qualities of tourism students. It has been a persistent issue in China's tourism education system that employers find new tourism graduates not ready to take the jobs needed. Yin and Meng (2018) noted that there has been a mismatch between students' career perceptions and employers' expectations and the industry's labour demand. With the constant evolvement of the industry, the expectations of the tourism workplace may also be changing. However, some general qualities of a tourism graduate may be commonly required. These may include qualities of problem-solving, teamwork and collaboration, emotional stability, international perspective, lifelong learning perspective, and professional attitude and compliance with professional code of ethics. In this regard, the TEFI (Sheldon et al., 2011) serves as an important reference framework. In consideration of China's education and social realities, graduate qualities of tourism students will be determined by industry demand and social expectations of tourism as a modern service industry. There is always a need to balance the vocational requirements with the liberal arts requirement in the curriculum design and delivery to enable the desirable graduate qualities. The relative weights on vocational skills and generic personnel qualities would be expected to vary across the level of the education provided, be it the secondary vocational education or tertiary research-oriented education.

It is expected that with the further transformation of the Chinese society and economy, economic restructuring may generate new tourism-related sectors and thus create new tourism-related job positions. For example, smart technologies have been increasingly applied in different areas of tourism in China. Accordingly, new jobs may be created around the applications of new technologies in tourism. If the industry develops fast to generate new jobs, the education system needs to be quick to pick up the demand and integrate the new market demand.

Internationalisation

Internationalisation has been a clear trend and development strategy of the international tourism education (Hobson, 2010; Hsu, 2017). Before the COVID-19 pandemic, China mainly served as a major student source for international tourism education providers. As elaborated in Chapter 10 by Luo and Zhai, different models of collaboration can be found between Chinese universities and foreign universities at both undergraduate and postgraduate levels. Student mobility is mainly from a Chinese university to a foreign country university rather than the other way around. Joint degree programs are at the core of such arrangements. Some international collaborations are above the program level, operating as joint schools or colleges. For example, Tianjin University of Commerce (TUC) has been collaborating with Florida International University (FIU) to operate its TUC-FIU Cooperation College; and Hainan University (HNU) and Arizona State University (ASU) jointly operate the HNU-ASU international Tourism College on the campus of HNU.

Many Chinese universities offering tourism programs still face the pressing need to strive for internationalisation. In addition to seeking joint degree programs and joint school operations, as mentioned above, many universities, even those less research-oriented universities, would request their teaching staff to have international visiting scholarship experiences. Some universities have clearly instituted "international visiting scholarship" experience in their staff promotion key performance indicators.

The COVID-19 pandemic and the changed geopolitical relations between China and many Western countries since the outbreak of the pandemic are likely to change the landscape of internationalisation on China's tourism education. Due to closed borders in many countries, student mobility has been greatly impacted. Students who intended to study aboard may eventually choose to study in China instead as they cannot enter the study destination country. The Chinese government has also issued warnings to students against some study destination countries due to increasing safety incidents reported in specific countries. These government warnings may divert students to choose other countries perceived to be safer.

In general, due to the combined effect of the pandemic and the new international relations driven by the US-China conflicts, there may emerge a declining trend for Chinese students to choose to study tourism in a foreign university. They may instead elect to study in a Sino-Foreign joint school operating in China or just choose to study in a Chinese university. Nevertheless, the perceived value of studying tourism in a foreign country university may have decreased substantially at the moment. Students may not find themselves any more competitive in the job market even with a tourism degree gained overseas.

Despite the expected declining number of students studying tourism in foreign countries, the need for Chinese universities to go international may only grow stronger. More international collaborations may be reflected in staff exchange and joint PhD programs, as many Chinese universities would join the tournament of international university ranking (e.g., ShanghaiRanking's academic subject ranking in Hospitality and Tourism Management) to increase their research outputs for a higher ranking position.

Another trend worth noting is that some Chinese universities, especially those highly ranked research-oriented universities, have already been operating international tourism study programs attracting foreign students. Most of these international study places have been offered to students from the Belt-and-Road countries in line with the Belt-and-Road national strategy. This trend will be further pumped by China's "soft-power" construction. As these international programs will be delivered in English as the language of instruction, there is a growing need for tourism PhD graduates who can teach in English. China seems to be joining the international chorus in offering qualifications in English (Hobson, 2010). In the future, China may become a more attractive international study destination for tourism studies if its economy sustains a more prosperous tourism industry.

Technological applications and digitisation

Before the pandemic, technological advancement and applications in the higher learning sectors had been a main driver of the transformation of higher education. Discussions and adoptions of new teaching and learning models in the higher education field, including massive open online course (MOOC) flipped learning, had been in place for many years. The COVID-19 pandemic accelerated the speed of development and applications of online teaching and learning technologies. A flexible learning environment supported by online streaming and meeting technologies seems to be an essential part of higher education provider in the new normal environment induced by the pandemic. Along the trend of digitisation and its multifaceted influences on every aspect of the society, it is understandable that the higher education system will also be greatly disrupted and transformed. Teachers' roles may be further redefined; some of the current teaching roles, such as marking and examinations, may be increasingly taken over by new technological applications, such as artificial intelligence (AI). Clearly, the traditional mode of face-to-face teaching delivery will be increasingly replaced by flexible, non-time-fixed ways of online teaching.

China has been increasingly adept in technological innovation and applications in various economic sectors and social aspects. During the pandemic, online teaching was quickly adopted by many Chinese universities with well-developed online meeting platforms. The pandemic, in effect, has accelerated the transformation of the higher education and learning modes. With many countries keeping their borders closed, physical mobility of students going across country borders to receive the desired higher education have been greatly impaired. Universities may need to reconsider their transnational student market and new models of program delivery to their international student market. It is expected that cross-national online and "fluid" degree provisions may become more popular (Hsu, 2018). Students may choose to stay in their home countries to learn a foreign degree through online delivery or move to different countries to complete the required components for a degree. An example of the latter is the Master of Science of Global Hospitality Business degree collectively offered by The Hong Kong Polytechnic University, University of Houston, and Ecole Hoteliere de Lausanne. Students in this program spend three semesters in three continents (Hong Kong, the US, and Switzerland) plus a final semester for a project. Students who have completed the program will receive a degree from their home institution and a certificate from the other two partners' institutions (Hsu, 2018).

With the aid of advanced online teaching and learning technology, possibly facilitated by virtual reality (VR) and augmented reality (AR) applications, future pervasion of technology into the higher learning environment will likely further transform the landscape of tourism education in China. It is likely more flexible learning modules in the form of micro-credentials will be preferred by both learners and education providers.

The development of educational technologies and their applications will also prompt tourism education providers to think and act more innovatively in the competitive environment. International collaborations will continue, but in response to future challenges of student mobility and cost of teaching delivery, more flexible forms of teaching and learning arrangement taking full use of online technologies may emerge. Those colleges, schools, and programs with more innovative and financially viable solutions may stand to last, while those unable to adapt may be squeezed out of the market.

Learning environment

With the applications of learning technologies also comes the transformation and shift of learning environment. Specific to tourism and hospitality education, the need to produce industry-ready graduates is more pronounced. As such, learning environment that integrates real workplace scenarios will be increasingly applied, possibly with the support of AI, AR, and VR technologies. For example, students studying tour guiding can practice their tour guiding commentary and interpretation skills in a simulated room with AR + AI prop-up for a virtual tour groups with AI-supported virtual tourists interacting and posting questions. With the increasing uses of service robots in the industry, tourism and hospitality students may also need to learn to work with service robots in delivering services to customers.

As predicted by Hsu (2018), traditional classrooms may become less effective in catering to the needs of future learners. Spaces of social collaborative learning, which create more flexible and engaging learning atmosphere, are on the rise and the distinction between formal and informal learning spaces is become increasingly blurred. Hsu (2018) elaborated on the concept of "hives" as a new learning environment. "Hives" represent learning hubs with activities and facilities for busy learners. These spaces can be used for both formal classes and informal social learning and will replace a lot of the current labs in the tourism education system.

There are many possibilities for an innovative, eye-opening, and learner-centred online virtual learning environment, combining the newest developments of AI, AR, and VR technologies. Such virtual learning environments may be more appealing to the tech-savvy millennial generation learners.

Status of tourism education

Tourism education in China has developed phenomenally and has emerged to be a significant part of the China's education system. Tourism has been regarded as one of the strategic pillar industries in China and is believed to be a key driver of the country's economic transition and development. Tourism education in China has formed a distinctive field and community of educators. The Tourism Education Association has been in operation for more than a decade under the auspice of China National Tourism Administration (Ministry of Culture and Tourism after 2018). And there are also various types of tourism education and teaching committees in different levels of higher education and vocational education system.

As mentioned in Chapter 1 and more or less hinted in other chapters of this book, there is a common belief among the community of tourism education in China that the status of tourism subject in China's higher education subjects system is low, and as such, the development of tourism as a university subject has been compromised. This issue is not unique in China. Even in universities in other countries, tourism, as a new subject area, does not enjoy an equal status with other traditional disciplines and subjects. Based on the nature of tourism as a field of study and research, the tourism subject is commonly hosted in the faculties of business, social sciences, and, to a lesser degree, humanities in a university. Compared to traditional disciplines like geography, history, economics, and management, tourism has a shorter period of development and thus does not enjoy a high regard by university administrators.

However, the situation may be gradually changing. In China, the "dual world class" subject construction campaign in the higher education system has already seen a significant number of Chinese universities listing "tourism management" or "hospitality management" as their flagship subjects. In the ShanghaiRanking's of Chinese university subjects, there are 262 universities ranked for the subject of "tourism management", 132 universities for "hotel management", 62 universities for "event economics and management", and 15 universities for "tourism management and services education".

As commented by Huang (2012), in a neoliberal development environment, the development of tourism as a university subject will be influenced by the many parametric factors of the institutional environment. One of such factors is exemplified by Huang (2012) as the national research assessment framework. Huang (2012) illustrated that the Excellence of Research in Australia (ERA), in contrast to the Research Assessment Exercise (RAE) in the UK, made tourism research in Australia more visible, as tourism is listed in both a two-digit field of research and a four-digit field of research descriptions. In effect, tourism as a subject and research field was treated more favourably in the Australian university system than in the UK university system.

Similarly, the ShanghaiRanking's subject ranking in Hospitality and Tourism management offers an institutional surveillance framework that can directly or indirectly guide stakeholders' attention to tourism as a university subject. Some universities in China, such as Sun Yat-sen University, Zhejiang University, Xiamen University, Fudan University, and Nankai University, have climbed up the ladder in this ranking list quickly in recent years. It is expected more Chinese universities will be entering this accolade list in the coming years, and once there are more universities in China in the top echelon positions of this ranking list, the Chinese tourism education community's bargaining power for more recognition of tourism as a university subject will grow stronger. As mentioned above, the ShanghaiRanking has also created "the best Chinese university subject ranking" following the Ministry of Education's Catalogue of Undergraduate Degree Subjects, in which Tourism Management is a first degree discipline under the discipline cluster of Management and comprises four subject areas of Tourism Management, Hotel Management, Event Economics and Management, and Tourism Management and Services Education. Such a ranking system of Chinese university subjects will further increase the visibility of tourism as a university subject in China. Accordingly, tourism as a university degree subject will likely gain more recognition in the Chinese university system and tourism education in China will grow to be more prominent internationally. Comparatively, tourism as a university subject may indeed enjoy a better status in China than in other countries.

Research

The ShanghaiRanking's university subject ranking in hospitality and tourism management will also promote the development of tourism research in Chinese universities. As shown in ShanghaiRanking's website (http://www.shanghairanking.com/), the global ranking of academic subjects relies predominantly on an institution's performance indicators in research. These indicators include a number of papers published in top quarter journals in the Web of Science database, the Category Normalized Citation Impact (CNCI) of the papers published by an institution in the Web of Science's InCites database, international collaboration, number of papers published in the top journals identified in each subject, and the recognised awards received. Almost all these five indicators showcase the performance of research over that of teaching. Therefore, ShanghaiRanking, although regarded as more objective in its ranking methodology, can be roughly treated as a research performance ranking among the major global university rankings.

China's public universities, especially those research-intensive universities, are generally well funded by the government. Chinese universities appear to be cash-rich compared to their counterparts in other countries. As a matter of fact, research funding in Chinese universities can be much better than that in foreign universities. Considering that Chinese universities are more self-sufficient in their student supply and less dependent on international students in their revenues, and that China has been dealing with the COVID-19 pandemic very efficiently and its universities are less affected financially, it is foreseeable that the gap between Chinese universities and their foreign counterparts in terms of research funding will be further enlarged. Bao, Huang, and Chen (2019) showed that since 2010, the grants awarded to tourism researchers in China had significantly increased, evidencing the increasing funding support of tourism research in China. In contrast, Huang, Yu, and Li (2019) found Australia's national government grants support to university-based tourism research has been marginal. With a prospect of better funding support, Chinese universities will produce more tourism research outputs, and possibly, the ShanghaiRanking's global ranking in the academic subject of hospitality and tourism management will see more Chinese universities moving up to the top positions in the future.

With more abundant funding support and the growing research capability of its researchers, Chinese universities will see its share of tourism research contributions to the international journals continuously growing. More productive and prominent authors will emerge from the tourism research community in China. As China provides a rich context for tourism research through its constantly evolving industry with myriad of innovative business models and industry practices, we also foresee that tourism research conducted by tourism researchers based in Chinese universities will increase its preponderance in international tourism research community. Of course, most of such research will continue to be produced through international collaborations in which authors in China work with their international co-authors to produce high-quality research outputs.

The changing landscape of tourism research in Chinese universities will also be transformed by the increasing number of PhD graduates overseas returning to China. As tourism has become a promising academic field for an academic career, more and more Chinese students have sought a doctoral degree in tourism or related fields in foreign country universities, most likely in universities in the US, the UK, Australia, New Zealand, and some European countries. On a global scale, tourism faculty positions in universities outside China are becoming limited and increasingly competitive. However, demand for teaching staff with PhD qualifications received overseas from the tourism and hospitality schools and programs in Chinese universities is strong. Many tourism PhD graduates, even those without a Chinese ethnic origin, may find China as a promising country to develop their academic career.

Further internationalisation of tourism faculty in Chinese universities will inevitably promote the international exchange of tourism scholarship and research between Chinese universities and their foreign counterparts. Although international relations between China and some Western countries will prove to be a roadblock for such research exchanges, the trend won't be affected much, as more foreign universities may find that it is to their own benefits to partner with Chinese universities for tourism research when they face more funding restrictions.

Institutional sustainability

Looking forward, it is expected tourism schools and colleges in Chinese universities will probably undergo a reshuffling process. During this process, strong players will grow stronger while some weak players may be squeezed out of the market. Two main factors among many related ones may determine the final results of the reshuffling. First, the market demand for tourism and hospitality education may be declining and consequently, universities will face more intense competition for student recruitment. If a university cannot recruit a sufficient number of students in its tourism and hospitality programs, these programs will be likely cut off for lack of financial viability. In the international tourism education market, program and school closures are not rare (Fidgeon, 2010). Modern universities normally have the "tripod" functions of teaching, research, and providing services to the society. If teaching cannot be maintained in a program, the program may be removed and faculty teaching in the program may need to be transferred to more financially sound programs. Second, global university rankings like the ShanghaiRanking will push many Chinese universities to invest resources and apply strategies to improve their research performance indicators. Research performance will drive up global ranking, which turns into reputation credits to attract more students in their tourism and hospitality programs. So the university ranking game will favour those research-oriented universities which can pull more of their staffing and funding resources into tourism research to maintain or strive for a high-ranking position nationally and internationally. As a result, these universities will maintain their competitive advantage in the market and thus can sustain their tourism and hospitality schools and programs.

The above-described reshuffling process may also happen with universities outside China that offer tourism and hospitality program. Before the COVID-19 pandemic, many universities in English-speaking countries such as Australia, UK, and US had become heavily reliant on international student markets in their tourism and hospitality programs. China was a key student source for these programs. The pandemic will have a long-term negative impact on the survival of these programs. Without sufficient student numbers, these programs will face the fate of closure. Despite this gloomy prospect of international education in tourism, China may still provide a hopeful future for international tourism education providers. As long as China maintains a healthy development in its economy and tourism remains to be a pillar industry sector to drive up domestic consumption, China's tourism industry will keep its vitality and require qualified employees from the global tourism education system. Universities outside China which offer tourism programs may still take a share of this growing market with appropriate strategies and in collaboration with key partners in China. In this regard, the past may not provide much clue for the future. Innovative business models and partnership operations beyond pre-pandemic practices are needed to deal with the "new-normal" in the post-pandemic era. Innovative teaching and learning activities utilising technological advancements in AI, VR, and AR will likely be the core of such models. It is predicted the role of research will be strengthened in the tourism education systems both within and outside China. On the one hand, research will be more prominent as a university function as the industry may increasingly resort to research to guide its development in an uncertain environment; on the other hand, and more importantly, tourism research in universities will increasingly lever the institutional reputation in tourism education and thus play a bigger role in sustaining tourism and hospitality programs.

Conclusion

This chapter concludes the whole book by providing discussions on the critical issues and challenges facing the tourism education system in China. In addition, this chapter also offers speculative comments on the future of tourism education development both within and outside China. Altogether, ten broad issues are discussed in this chapter, covering student enrolment scales, curriculum, teachers, students, internationalisation, technological applications, future learning environment, status of tourism education, and research and institutional sustainability of tourism programs. The future of tourism education inside and outside China may be beyond our imagination in the long term. This chapter, nevertheless, provides some glimpses into the future based on the authors' observations and experiences with both the tourism industry and tourism education sectors in and outside China.

References

Ayikoru, M., Tribe, J., & Airey, D. (2009). Reading tourism education: Neoliberalism unveiled. *Annals of Tourism Research, 36*(2), 191–211.

Bao, J., Huang, S., & Chen, G. (2019). Forty years of China tourism research: Reflections and prospects. *Journal of China Tourism Research, 15*(3), 283–294.

Dredge, D., Benckendorff, P., Day, M., Gross, M.J., Walo, M., Weeks, P., & Whitelaw, P. (2012). The philosophic practitioner and the curriculum space. *Annals of Tourism Research, 39*(4), 2154–2176.

Fidgeon, P. (2010). Tourism education and curriculum design: A time for consolidation and review? *Tourism Management, 31*(6), 699–722.

Hobson, J.S.P. (2010). Ten trends impacting international hospitality and tourism education. *Journal of Hospitality and Tourism Education, 22*(1), 4–7.

Hsu, C.H.C. (2017). Internationalisation of tourism education. In P. Benckendorff & A. Zehrer (Eds.), *Handbook of teaching and learning in tourism* (pp. 321–335). Cheltenham: Edward Elgar.

Hsu, C.H.C. (2018). Tourism education on and beyond the horizon. *Tourism Management Perspectives, 25*, 181–183.

Huang, S. (2012). Similar exercises, different consequences: An examination of tourism research in national research assessment frameworks. *Tourism Management Perspectives, 2–3*, 13–18.

Huang, S., Yu, Z., & Li, Z. (2019). Australia's national government grants in tourism research. *Anatolia, 30*(4), 629–631.

Sheldon, P.J., Fesenmaier, D.R., Tribe, J. (2011). The Tourism Education Futures Initiative (TEFI): Activating change in tourism education. *Journal of Teaching in Travel & Tourism, 11*(1), 2–23.

Tribe, J. (2006). The truth about tourism. *Annals of Tourism Research, 33*(2), 360–371.

Yin, Z., & Meng, F. (2018). Tourism higher education in China: Profile and issues. In J. Zhao (Ed.), *The hospitality and tourism industry in China: New growth, trends, and developments* (pp. 241–261). Palm Bay, FL: Apple Academic Press.

Yu, C., & Zeng, G. (2016). Personnel training and industry demand on tourism management discipline. *Tourism Tribune, 31*(10), 18–19.

Zhang, W., & Fan, X. (2005). Tourism higher education in China. *Journal of Teaching in Travel & Tourism, 5*(1–2), 117–135.

Zins, A.H., & Jang, S.Y. (2019). Review and assessment of academic tourism and hospitality programmes in China. In C. Liu & H. Schänzel (Eds.), *Tourism education and Asia* (pp. 81–105). Singapore: Springer Nature Singapore.

Index

For Product Safety Concerns and Information please contact our EU
representative GPSR@taylorandfrancis.com
Taylor & Francis Verlag GmbH, Kaufingerstraße 24, 80331 München, Germany